Environmental Economics and Investment Assessment III

WIT*PRESS*

WIT Press publishes leading books in Science and Technology.
Visit our website for the current list of titles.
www.witpress.com

WIT*eLibrary*

Home of the Transactions of the Wessex Institute.
Papers presented at Environmental Economics III are archived in the WIT eLibrary in volume 131 of WIT Transactions on Ecology and the Environment (ISSN 1743-3541).
The WIT eLibrary provides the international scientific community with immediate and permanent access to individual papers presented at WIT conferences.
Visit the WIT eLibrary at www.witpress.com.

THIRD INTERNATIONAL CONFERENCE ON
ENVIRONMENTAL ECONOMICS
AND INVESTMENT ASSESSMENT

Environmental Economics III

CONFERENCE CHAIRMEN

K. Aravossis
National Technical University of Athens, Greece

C.A. Brebbia
Wessex Institute of Technology, UK

INTERNATIONAL SCIENTIFIC ADVISORY COMMITTEE

Organised by
Wessex Institute of Technology, UK
The National Technical University of Athens, Greece

Sponsored by
WIT Transactions on Ecology and the Environment

WIT Transactions

Transactions Editor

Carlos Brebbia
Wessex Institute of Technology
Ashurst Lodge, Ashurst
Southampton SO40 7AA, UK
Email: carlos@wessex.ac.uk

Editorial Board

Environmental Economics and Investment Assessment III

Editors

K. Aravossis
National Technical University of Athens, Greece

&

C.A. Brebbia
Wessex Institute of Technology, UK

WIT*PRESS* **Southampton, Boston**

K. Aravossis
National Technical University of Athens, Greece

C.A. Brebbia
Wessex Institute of Technology, UK

Published by

WIT Press
Ashurst Lodge, Ashurst, Southampton, SO40 7AA, UK
Tel: 44 (0) 238 029 3223; Fax: 44 (0) 238 029 2853
E-Mail: witpress@witpress.com
http://www.witpress.com

For USA, Canada and Mexico

Computational Mechanics Inc
25 Bridge Street, Billerica, MA 01821, USA
Tel: 978 667 5841; Fax: 978 667 7582
E-Mail: infousa@witpress.com
http://www.witpress.com

British Library Cataloguing-in-Publication Data

A Catalogue record for this book is available
from the British Library

ISBN: 978-1-84564-436-9
ISSN: 1746-448X (print)
ISSN: 1743-3541 (online)

The texts of the papers in this volume were set individually by the authors or under their supervision. Only minor corrections to the text may have been carried out by the publisher.

Printed in Great Britain by MPG Books Group, Bodmin and King's Lynn.

Preface

This book contains the edited versions of papers presented at the Third International Conference on Environmental Economics and Investment Assessment, held in Cyprus, in 2010. The conference was organised by the Wessex Institute of Technology in collaboration with the National Technical University of Athens.

The current emphasis on sustainable development is a consequence of the general awareness of the need to solve numerous environmental problems resulting from our modern society. This has resulted in the need to assess the impact of economic investments on the environment. The book addresses the topic of Investment Assessment and Environmental Economics in an integrated way; in accordance with the principles of sustainability considering the social and environmental impacts of new investments.

The usual assumption is that it is difficult to achieve the growth of enterprise in an environmentally friendly manner. This paradigm usually associated with developed countries is now affecting all other regions of the globe. The main question is if the development of enterprise is compatible with environmental protection.

The roots of financial development are financial growth, which in conventional terms requires an increase in production and the use of more resources. Overuse of those resources can result in ecological destruction and a larger release of waste and pollution into the environment.

The book addresses these problems of primary importance to Society, discussing and proposing a more constructive and progressive approach to ensure sustainability. Methodologies to address these important problems are presented.

The contributions comprise the following broad subject headings: Environmental Policies, Planning and Assessment; Cost Benefits Analysis; Decision Support Systems; Natural Resources Management; Social Issues and Environmental Policies.

This book will be of interest to government officials, politicians, environmental experts, economists, research scientists in the area of environmental economics, operations researchers, senior management in all kinds of companies and regional government.

The Editors are grateful to all the authors for their excellent contributions and in particular to the members of the International Scientific Advisory committee as well as all reviewers for their reviews of the abstracts and the papers and their help on ensuring the high quality of this book.

The Editors
Cyprus, 2010

Contents

Section 2: Cost benefits analysis

Section 3: Decision support systems

Section 4: Natural resources management

Section 5: Social issues and environmental policies

Section 1
Environmental policies, planning and assessment

The impact of the emission trading system on companies' profitability: the case of Greece

K. Aravossis & G. Garoufi
Sector of Industrial Management & Operational Research,
School of Mechanical Engineering,
National Technical University of Athens, Greece

Abstract

The European Union is in many respects a key player in the global efforts to curb emissions of greenhouse gases (GHGs). Maintaining the role of the frontrunner, the European Parliament and the Council established on 13 October 2003 a scheme of GHG emission allowance trading within the Community, the so-called Emission Trading System (ETS).

Greece, as a Member State of the European Union, takes action in the field of the reduction of carbon dioxide (CO_2) emissions. According to the Directive 2003/87/EC, the Greek government includes in National Allocation Plans (NAPs) the biggest polluters from each one of the energy demand sectors. More specifically, the Directive covers electricity industries, other industrial combustion installations, refineries, metal ore roasting and cindering installations, pig iron and steel production installations, cement clinker production installations, lime production installations, glass manufacture installations, ceramic production installations and pulp and paper production installations.

The objective of this paper is to examine the impact of the Emission Trading System on the profitability of Greek companies included in the NAPs. On this basis, the quantities of carbon dioxide that every participant has emitted during each year of the period 2005-2008 are compared respectively to the quantity of emission allowances issued. The balance indicates whether the tradable allowances are responsible for the participants' financial results (of their balance sheets).

WIT Transactions on Ecology and the Environment, Vol 131, © 2010 WIT Press
www.witpress.com, ISSN 1743-3541 (on-line)
doi:10.2495/EEIA100011

Financial indicators are used to present the impact of the acquisition or sale of allowances on the total turnover and the profit before tax of the participating companies.

An additional goal of this paper is to identify the causes that resulted in a surplus or shortage of allowances, taking into consideration financial, as well as environmental, parameters. Consequently, some useful conclusions are drawn concerning the management of allowances that are granted for free to the participating companies.

Keywords: greenhouse gases, CO_2 emissions, emission trading system, allowances, companies' profitability.

1 Introduction

The United Nations Framework Convention on Climate Change (UNFCCC) and the Integrated Pollution Prevention and Control (IPPC) Directive were the predecessors of the Kyoto Protocol, which sets targets, timetables and legally binding emissions, for Annex I countries, concerning carbon dioxide (CO_2), and other greenhouse gases. The foundation stone and the legal bedrock of the flexible market mechanisms were laid by the UNFCCC, but the Kyoto Protocol established them, in order to coordinate the efforts in the field of cost-effective environmental protection. Directive 2003/87/EC of the European Parliament and the Council established a scheme of greenhouse gas emission allowance trading within the community, the so-called Emission Trading System (ETS) [1].

In Greece, the activities covered by the provisions of Directive 2003/87/EC differ significantly in terms of the share of their emissions from combustions and emissions from processes. Considering this, the allocation of emission allowances is carried out in two stages, first at activity level and then at installation level. The activities under examination are the ones referred in Annex I of the Directive, namely energy activities and other combustion installations, mineral oil refineries, production and processing of ferrous metals, production of cement and lime, manufacture of glass and ceramic products and production of paper and cardboard [2].

2 The Greek national allocation plans

The first Greek NAP, concerning the three-year period 2005-2007, included 141 installations and allocated them approximately 223,3 Mtn CO_2 [3], while the second Greek NAP, concerning the first period of commitment under the Protocol, 2008-2012, includes 140 installations and allocates them approximately 341,5 Mtn CO_2 [4]. The installations that finally submitted verified emission reports for the first period outnumbered those initially included in the NAP. On the contrary, the installations that submitted annual verified emission reports for 2008, the first year of the five-year period, were noticeably less than those which were initially included in the second NAP. This reduction depends to a great extent on the financial crisis which led to malfunction or even closure of a considerable number of other combustion installations.

3 The impact of the ETS on Greek companies' profitability

3.1 The trends of CO_2 emission allowance prices

Trading of carbon dioxide emissions takes place in marketplaces since 2005. During the first year of operation of the multi-country, multi-sector GHG Emission Trading System, EUAs (EU Allowance Unit of one tonne of CO_2) were priced on the average at 21.37 €/tn CO_2 eq. [5], after many fluctuations, by reason of immaturity of the market.

In 2006, the market was, once more, characterized by instability and EUAs were priced at 18.18 €/tn CO_2 eq. on the average. Prices decreased, due to the fact that EUAs for trading period 2008-2012 stimulated a lot of interest.

In 2007, EUA prices plummeted. The price of a tonne of carbon dioxide in Europe often fell below 1 €/tn CO_2 eq. [5]. Demand for permits to emit CO_2 dropped off, since traders had lost all interest of first trading period's EUAs and the price was 1.44 €/ton CO_2 eq. on the average.

Prices were kept stable during 2008. In grace of traders' raised interest, EUAs were priced at 22.66 €/tn CO_2 eq. [5] on the average.

3.2 The impact of the ETS through diagrams and financial indicators

Undoubtedly, the ETS has influenced Greek companies' profitability. The average emission allowance price per year, as well as data concerning the companies' CO_2 emissions [6] were required in order to examine thoroughly its impact.

The following diagrams present the income from the surplus CO_2 emission allowances and the expenses on the acquisition of extra ones, of representative companies from each sector.

Moreover, two financial indicators were used so as to examine the financial situation of the above companies after the installation of the ETS. The impact of the acquisition or sale of CO_2 emission allowances on the total turnover and the profit before tax of the participating companies was calculated, using financial results from the companies' balance sheets.

3.2.1 The impact of the ETS on the total turnover of the participating sectors' companies

The first diagram depicts the comparison between the average of indicators presenting the impact of the acquisition or sale of emission allowances on the total turnover of the companies during 2005-2007 and 2008. More specifically, the first bar concerns 2008, while the second one concerns the three-year period 2005-2007.

By studying the diagram, it is obvious that the ETS doesn't exert much influence on the paper and cement clinker production sectors.

Nonetheless, the lime and ceramic production installations had either shortage or small surplus of allowances during the first tradable period, whereas big part of their income proceeds from the sale of the surplus CO_2 emission allowances in 2008.

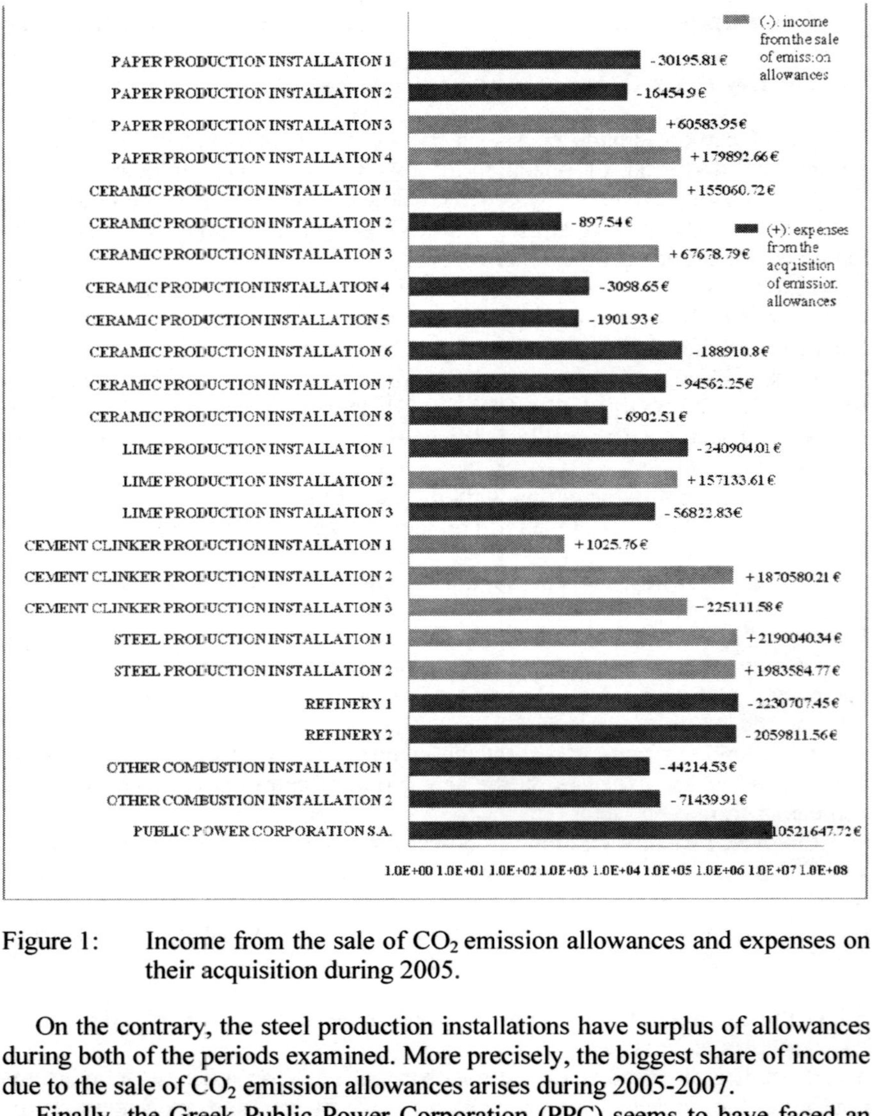

Figure 1: Income from the sale of CO_2 emission allowances and expenses on their acquisition during 2005.

On the contrary, the steel production installations have surplus of allowances during both of the periods examined. More precisely, the biggest share of income due to the sale of CO_2 emission allowances arises during 2005-2007.

Finally, the Greek Public Power Corporation (PPC) seems to have faced an extended shortage of allowances in 2008.

3.2.2 The impact of the ETS on the profit before tax of the participating sectors' companies

The second diagram pictures the comparison between the average of indicators presenting the impact of the acquisition or sale of emission allowances on the profit before tax of the companies during 2005-2007 and 2008. Likewise, the

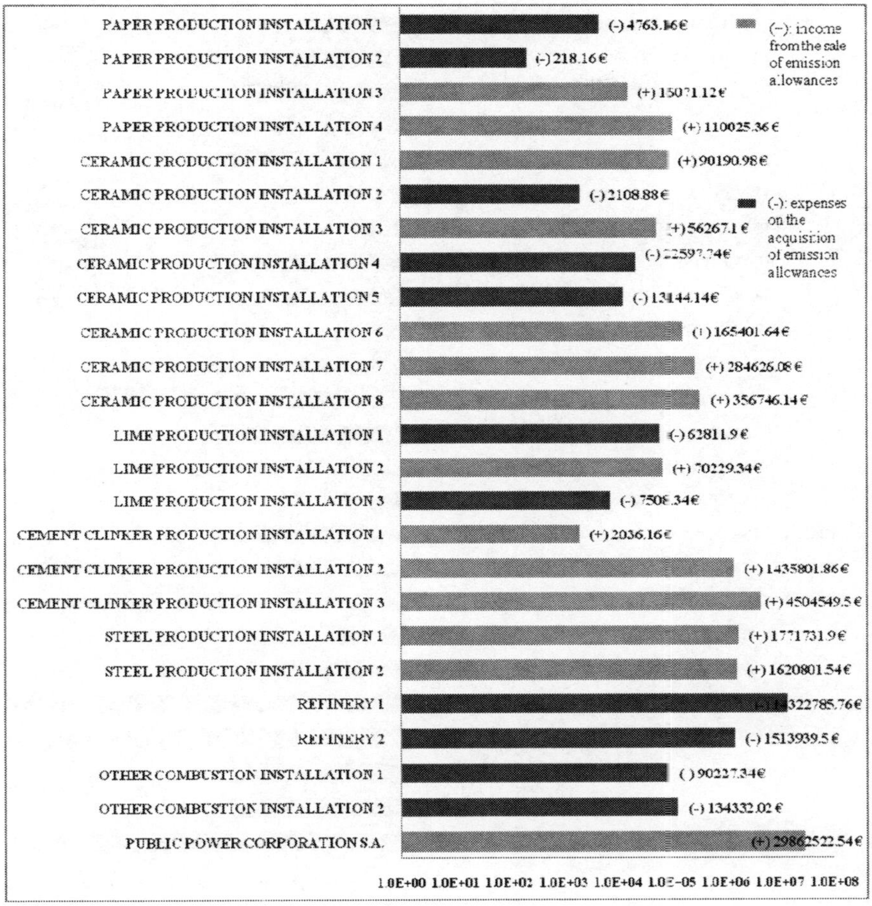

Figure 2: Income from the sale of CO_2 emission allowances and expenses on their acquisition during 2006.

first bar concerns 2008, while the second one concerns the three-year period 2005-2007.

Undoubtedly, the ETS has aided numerous companies from the lime and ceramic production sector to improve their economic situation, especially since the beginning of the financial crisis.

Regarding the steel production installations, the diagram gives the impression that the ETS has helped them raise their profits.

Oppositely to the above, the ETS unquestionably presides over 45% of the heavy loss the PPC announced in 2008.

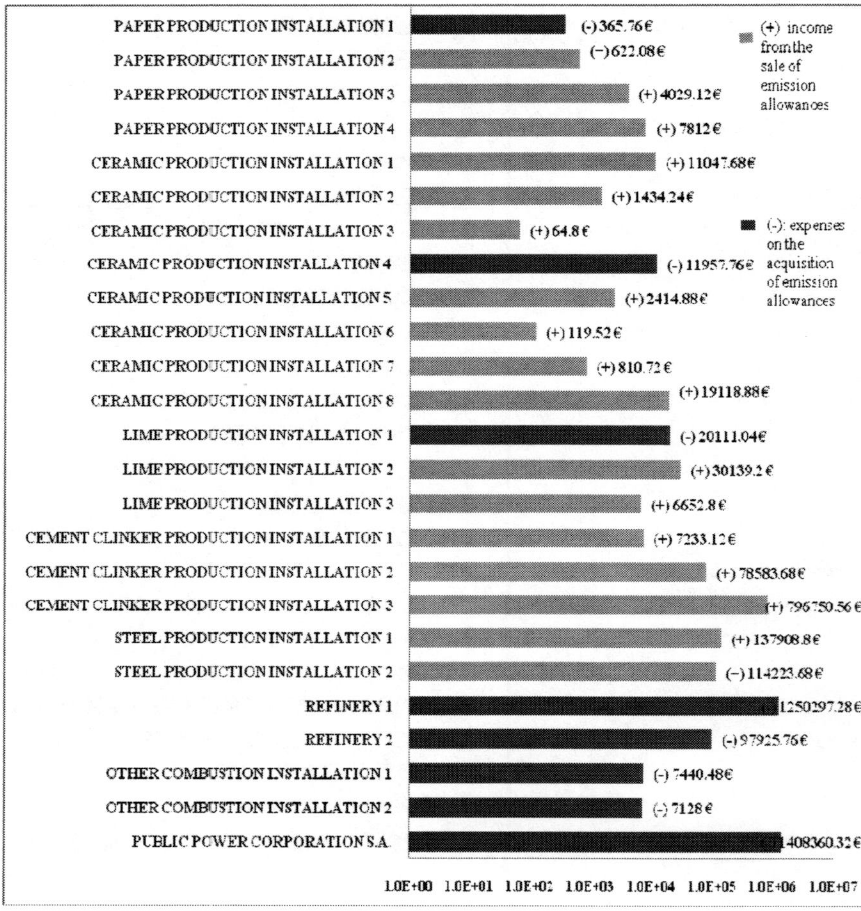

Figure 3: Income from the sale of CO_2 emission allowances and expenses on their acquisition during 2007.

4 Analysis of the impact of the ETS on Greek companies

4.1 Power generation sector

As far as the Greek Public Power Corporation (PPC) is concerned, the ETS seems to incur much of the damage in the company's profitability. Not only has the PPC shouldered the responsibility to cover the energy demands of the majority of Greek capitals, but also operates technologically old electricity generation units, using lignite [7]. The above factors explain the company's shortage of emission allowances, which is growing in 2008, since it's the first year of a period that demands reduction of CO_2 emissions.

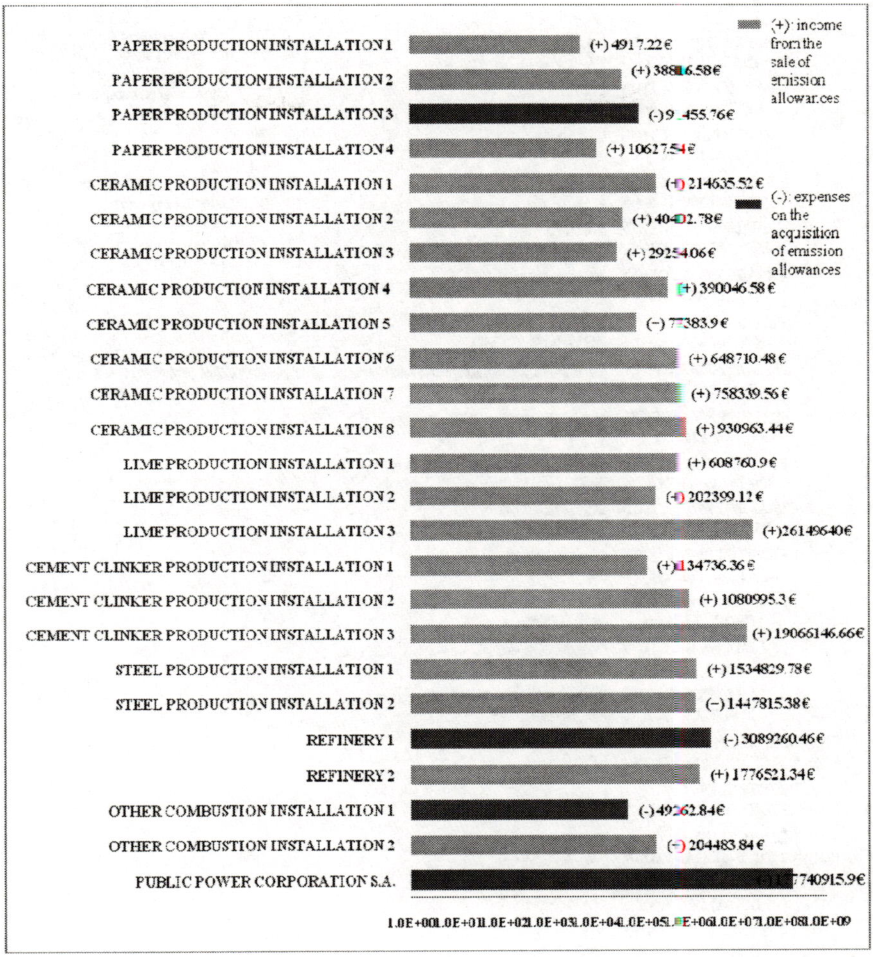

Figure 4: Income from the sale of CO_2 emission allowances and expenses on their acquisition during 2008.

Electricity market liberalization led to the entrance of three new units. Their parent companies were aware of the establishment of the ETS in advance, therefore, they had provided their subsidiary companies with combined cycle natural-gas units [8]. Hence, carbon dioxide emitted by them didn't outnumber the quantity of emission allowances issued.

4.2 Other combustion installations sector

The sector of other combustion installations, generally, presents limited CO_2 emissions during 2008, compared respectively to those of the three-year period 2005-2007 [6].

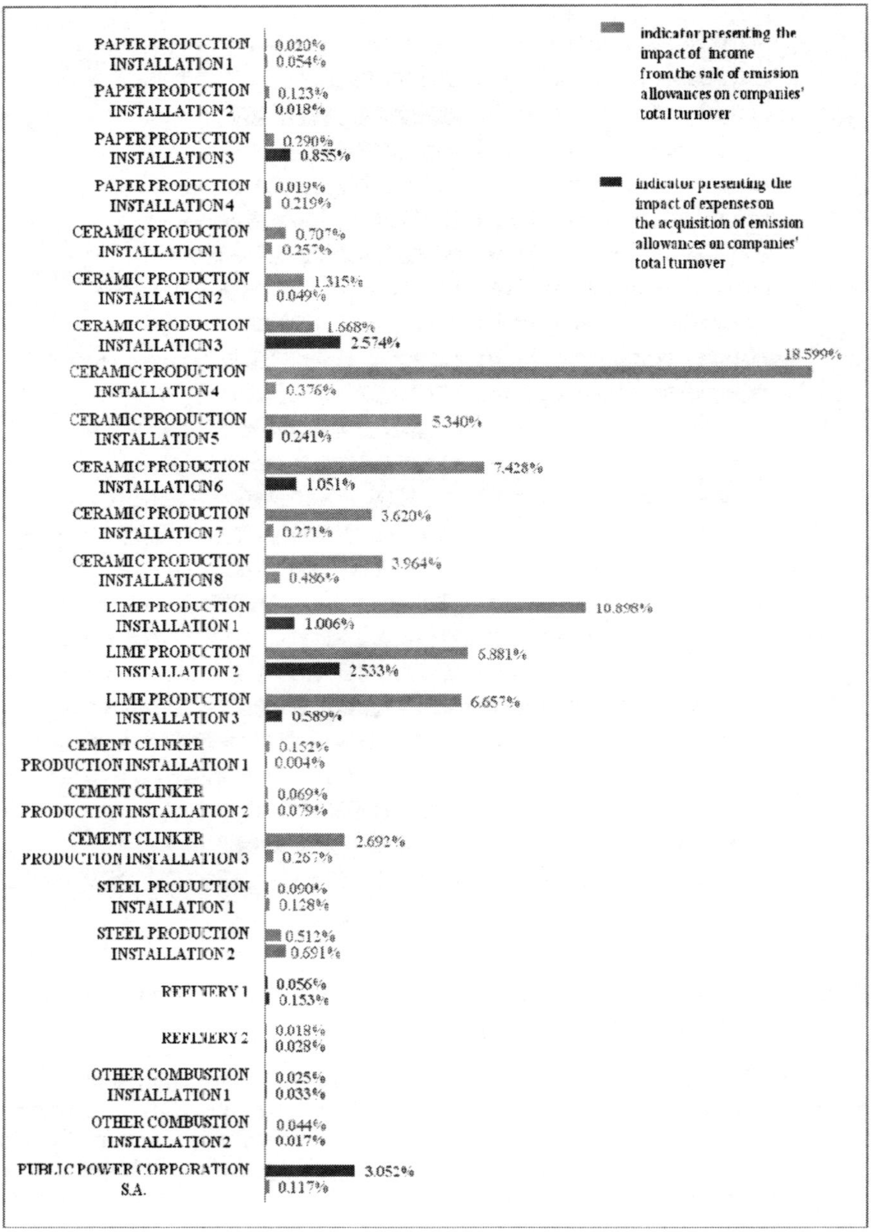

Figure 5: Comparison between the average of indicators presenting the impact of the acquisition or sale of emission allowances on companies' total turnover during 2005-2007 and 2008.

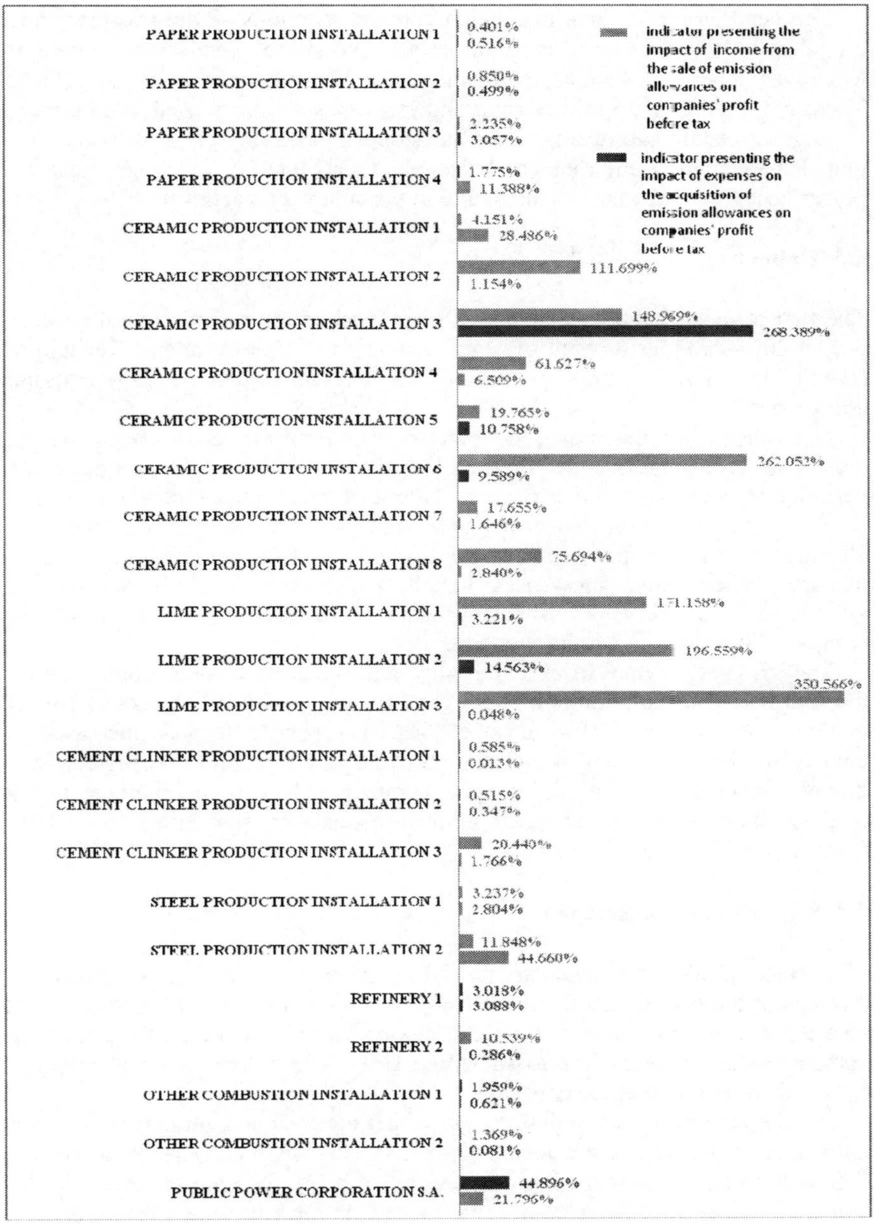

Figure 6: Comparison between the average of indicators presenting the impact of the acquisition or sale of emission allowances on companies' profit before tax during 2005-2007 and 2008.

The fact that the installations included in the second NAP are less than those included in the first one, brought about considerable emission reductions. Moreover, three installations malfunction and two others have already closed. Therefore, their quantity of emission allowances issued is not used. Furthermore, the biggest installation of this sector has stopped electricity generation activities that demanded non environmental friendly fuels. Instead, combined heat and power, using natural gas, was integrated in the company's activities [9].

4.3 Refineries

The first refinery has progressively reduced its shortage of emission allowances during 2005-2007, as a result of implementation of Best Available Techniques (BAT) [10]. These actions contributed directly in surplus of CO_2 emission allowances in 2008.

Notwithstanding the above, the total allowances of the sector were increased comparatively to the quantity of emission allowances issued for the three-year period, due to an expansion of the installations of the sector's second refinery.

The expansion took place in 2006, but no additional allowances were allocated to the company. Thus, the quantity of CO_2 emitted exceeded the quantity of emission allowances issued, which led to a big shortage of allowances. Therefore, a respectable amount of money was spent by the company, in order to cover its shortage.

The assigned amount of emission allowances for the second refinery during the first period of commitment under the Protocol (2008-2012), was calculated on the basis of its historical emissions. Thereupon, it was increased in comparison to the amount of emission allowances of the previous period. Even though the shortage was diminished, the company still spent much money to tide over the shortage of CO_2 emission allowances, as a consequence of high EUA prices during 2008 [5].

4.4 Steel production sector

The steel production installations have surplus of emission allowances throughout the period examined. The participating companies of this sector were aware of the establishment of the ETS. Taking into consideration the above, the companies had enough time to substitute HFO with natural gas and therefore reduce their carbon dioxide emissions [11].

Besides, the historical emissions, on which the assigned amount of emission allowances is based, were calculated while the installations were using HFO, a fuel that causes extremely high emissions of CO_2. Wherefore, the fact that allocated emission allowances were far more than the allowances the companies really necessitated, had an enormously positive impact on their profitability.

As far as the second NAP is concerned, substitution of HFO with natural gas was taken under consideration and as a result, allocated emission allowances were significantly less than the ones of the previous period.

4.5 Construction sector

The construction sector, on the whole, presents reduced CO_2 emissions compared to those of the three-year period 2005-2007. The financial crisis led the sector to a slowdown, and as a result the demand diminished [12]. In consequence of the limited demand, there was a proportionally limited production which restrained CO_2 emissions in low levels.

Additionally, the staple of the installations included in the construction sector (cement clinker production installations, lime production installations, ceramic production installations) is limestone. Thereupon, all the relative installations have the ability to use limestone of low percentage of CO_2 content and consequently reduce their emissions.

4.5.1 Cement clinker production sector

Regarding the cement clinker production sector, especially during the three-year period 2005-2007, there is no remarkable variance between the quantity of emission allowances issued and the quantity of CO_2 emitted. More specifically, the quantities of carbon dioxide emitted that are presented in the submitted verified emission reports, are slightly less than the allocated emission allowances, due to the following factors.

First and foremost, the cement clinker production sector is characterized by stability. It is consisted of a small number of companies and as a result, demanded quantities of cement, which indicate the produced quantities of cement and consequently the quantity of CO_2 emitted, are approximately stable from year to year. On account of that, historical emissions, on which calculations of the allocated emission allowances were based, suggest accurately the sector's demand for emission allowances.

On the other hand, participating companies of this sector, were beforehand aware of the establishment of the ETS and substituted fossil fuels with natural gas, which releases much less quantities of CO_2 [11].

Particularly in 2008, Greek cement clinker production companies confirmed reduction of carbon dioxide emissions up to 5%. This reduction emanated from the slowdown of the construction sector, due to the recent financial crisis. In 2008, the three biggest cement associations acted according to their estimation that downturn would continue, and sold big amounts of allowances through international marketplaces, in order to support their economic situation [6].

4.5.2 Lime and ceramic production sectors

Remarkable similarities can be identified between the lime and ceramic production sectors.

To begin with, none of the relative installations has renewed its equipment, or substituted fossil fuels with natural gas, since the establishment of the ETS.

What is more, during the period 2005-2007, allocated emission allowances proved to be insufficient to cover the needs of the installations included in the NAP. Thus, the quantity of CO_2 emitted exceeded the quantity of emission allowances issued, which led to a big shortage. As a consequence, the companies spent considerable amounts of money so as to tide over the shortage [12].

Under these circumstances, the second NAP allocated more emission allowances to the above companies for the period 2008-2012. However, the extra allowances that were at the companies' disposal turned out to be needless, since their verified emission reports of 2008 present significantly reduced emissions of carbon dioxide, owing to two main factors.

First of all, the financial crisis impacted on the construction sector, and by extension on the lime and cement clinker sectors, leading to their slowdown. As a result, the emerging CO_2 emissions dropped [6].

Additionally, several companies adopted the policy of further reduction of their production, aiming at the creation of a surplus of allowances. The sale of the surplus of allowances enabled them to support their economic situation.

4.6 Paper production sector

The paper production installations present slight differences between the quantity of emission allowances issued and the quantity of CO_2 emitted. With regard to this sector, paper production is based exclusively on recycled paper, and not on wood pulp (wood pulp is usually bleached using calcium carbonate, in order to produce white paper product). Hence, CO_2 emissions result only from combustion.

During 2008, the decision several paper production installations made to replace outdated technology, led to reduction of carbon dioxide emissions [13]. Some of them, having access to natural gas pipelines, have substituted fossil fuels with natural gas. Others use LPG either exclusively or in the mix of fuels [11].

5 Conclusions

This paper reported the impact of the Emission Trading System on Greek companies' profitability and its influence on their environmental policy.

The ETS, being responsible for the emergence of a big shortage of allowances and by extension of money, has worsened the economic situation of the Greek Public Power Corporation. More accurately, it unquestionably presides over the 45% of the heavy loss the PPC announced in 2008. Yet, the company hasn't taken sufficient measures to reduce CO_2 emissions, and instead operates mainly lignite stations.

Nevertheless, the other combustion installations and the paper production installations have remained almost unaffected. The first ones don't exceed the quantity of emission allowances due to their malfunction or even closure. The majority of the second ones have substituted fossil fuels with environmental friendly fuels and replaced outdated technology.

The sector of refineries was affected by the ETS, due to an expansion of one of the installations, during the first period examined, without being allocated additional emission allowances. Even when the second NAP took the expansion under consideration, CO_2 emission allowance prices were so high that loss was unavoidable.

The fact that the companies from cement clinker and steel production sectors were beforehand aware of the establishment of the ETS, in combination with the financial crisis, led to limited CO_2 emissions. As a result, they took advantage of the current situation, sold big amounts of allowances and increased their income.

The lime and ceramic production sectors were, indubitably, most aided by the ETS. CO_2 emission allowances became the needed source of income which helped them improve their economic situation, or even avoid their closure.

Beyond shadow of doubt, the Emission Trading System has underlined the importance of using Best Available Techniques and substituting fossil fuels with environmental friendly ones, like natural gas.

To sum up, the Kyoto Protocol, despite its apparent flaws in its current form, is the first international environmental agreement that sets legally binding GHG emissions targets and timetables for Annex I countries. If properly designed, emission trading scheme can effectively reduce their abatement costs while assisting Annex I countries in achieving their Kyoto obligations and assuring their economic viability.

References

[1] Directive 2003/87/EC of the European Parliament and of the Council of 13 October 2003 establishing a scheme for greenhouse gas emission allowance. Trading within the Community and amending Council Directive 96/61/EC.

[2] Sioulas K., Applying Emissions Trading Mechanisms in Greece, CRES, pp. 7, 2006.

[3] Ministry of Environment, Energy and Climate Change, National Allocation Plan for the period 2005-2007, pp. 19-26, 2004.

[4] Ministry of Environment, Energy and Climate Change, National Allocation Plan for the period 2008-2012, pp. 6- 9, 2008.

[5] European Climate Exchange, www.ecx.eu

[6] Kourniotis S., Tsouma M., Sellas N., Loukatos A., Gargoulas N. & Koryzi K., Verified emission reports of 2008 & Energy saving in businesses. *CO_2NTROL info*, EPEM, 13, pp. 6-14, 2009.

[7] Regulatory Authority for Energy, www.rae.gr

[8] Kourniotis S., Loukatos A., Gargoulas N., Mentzis A., Sitara A. & Makrinou K., 2nd Report concerning the establishment of the Emission Trading System. *CO2ntrol info*, EPEM, 2, pp. 12-13, 2006.

[9] Ministry of Development, Energy Outlook of Greece, pp. 63-68, 2009

[10] Hellenic Network for Corporate Social Responsibility, www.csrhellas.org

[11] Kourniotis S., Loukatos A., Gargoulas N., Sellas N., Sitara A., Mentzis A.& Makrinou K., Report of CO2 emission allowances. *CO_2NTROL info*, EPEM, 6, pp. 10, 2007.

[12] Construction activity as a parameter of economic development and the consequences of the economic crisis on it (Initiative Opinion); Economic and Social Council of Greece, Opinions, 225. http://www.oke.gr/oke_pron_pdf_en.html

[13] National Centre for the Environment and Sustainable Development, http://www.ekpaa.greekregistry.eu/

Infrastructure and ecology: 'limited' costs may hide substantial impacts

E. J. Bos[1] & J. M. Vleugel[2]
[1]*LEI Wageningen University and Research Centre, The Netherlands*
[2]*OTB Research Institute, TU Delft, The Netherlands*

Abstract

In response to a growing demand for transport and changes in the way people use space, the nature of road infrastructure networks change: (small) roads are frequently transformed into highways. Before such expansions are realized, in many cases a legal obligation exists to carry out a cost-benefit analysis (CBA) in order to assess whether the expansion scheme creates a net social benefit for society. A CBA deals with the effects on the surrounding ecosystem, the environment and human living. This paper focuses on the valuation of the biotic, a-biotic and socio-economic damage, thereby contributing to the methodology of CBA as a tool to evaluate infrastructural plans integrally.
Keywords: infrastructure, ecological effects, environmental effects, economic valuation and CBA.

1 Introduction

Cost-benefit analysis (CBA) has become an important tool to support policy making on public investments in infrastructure. Here we will focus on assessing the external effects of infrastructure in economic terms. More specifically, we will value the ecological and environmental effects of transforming an existing motorway into a highway.

The set-up of the paper is as follows. Section 2 starts with an introduction into CBA. Section 3 continues with an assessment of the ecological, environmental and socio-economic impacts of the highway plan, followed by the economic valuation of these impacts. Then an alternative plan to reroute the highway will be discussed briefly. Section 4 follows with an evaluation of the previous analysis. In section 5 the main conclusions and recommendations can be found.

WIT Transactions on Ecology and the Environment, Vol 131, © 2010 WIT Press
www.witpress.com, ISSN 1743-3541 (on-line)
doi:10.2495/EEIA100021

2 Methodology

Cost Benefit Analysis (CBA) is a well-known tool to support policy makers in making decisions about public investments in infrastructure- and other projects.

In this study, we undertook an extensive review of existing ecological and environmental studies in order to determine the physical impacts of a highway on nature and the environment. This provided a set of parameters. We combined these parameters with data from the plan in order to determine the quantitative impacts of the plan. The translation of these impacts in monetary terms, better known as economic valuation, is the last step to determine the social costs and benefits of the project. It contains uncertain elements, in particular regarding ecological effects [1]. There is also no straightforward, simple and integrative method to value ecological, environmental and socio-economic impacts, instead we combined different valuation methods in the assessment. To some extent this meant navigating at the edge of present knowledge, hence the indicative nature of our final results.

3 The plan for the highway and its impacts

3.1 Introduction

Kresna Gorge is a small area (17 km in length) located north of the Bulgarian town of Kresna in the southeast part of Bulgaria near the Greek border. Located on the border of the continental and Mediterranean climate zones, the gorge contains a unique ecosystem with a high biodiversity and rare animals and plants: a corridor for mammals and birds.

Nearly 5%, including the most valuable habitats, belongs to the protected Tissata Reserve. The gorge is declared as a CORINE site (Bern Convention) and will become part of the European Ecological Network Emerald and Natura 2000. Part of the gorge is also defined as an important area according to Bird Life

Figure 1: Kresna Gorge in Bulgaria [2, 3].

International criteria. Bulgarian environmentalists aim to turn the whole Kresna Gorge into a protected area. It could be linked with FYROM's nature areas as a Transborder Nature Park [4].

3.2 The existing motorway

The gorge and Kresna town are divided by the nine meter wide international road between Sofia and Athens (Figure 2). Traffic on this road kills hundreds of animals, such as (rare) snakes, polecats, tortoises, bats and otters, during their daily migration to the Struma River [4, 5].

3.3 The plan for the highway

Since 1997, a plan by the Bulgarian government did exist to replace the existing 2x1 lane motorway by a 2x2 lane (plus emergency lanes) E-79 highway. The European Union financially supports infrastructure projects of international importance by its Trans European Network (TEN) policy. The E-79 is part of Priority Project N° 7 (Igoumenitsa/Patra–Athina–Sofia–Budapest). An evaluation says [6, p. 18]: "Bulgaria intends to invest a major part of its Cohesion Fund 2007-2013 on the motorway route Sofia–Kulata (the *Struma* motorway). However, serious environmental constraints could lead to delays on a 56 km section at the "Kresna Gorge". These 'constraints' refer to the highway dissecting the Kresna Gorge and passing the edges of Kresna town at 30 meters.

3.4 The physical impact of the highway

When in use, the highway will produce a range of ecological, environmental and economic effects [4], which will be described here. Besides these external costs, the cost of building the highway is € 1.2 billion [7].

3.4.1 Ecological (biotic) impact

The presence of a road alters hydrological dynamics, disrupts natural processes and habitats, but may also create new habitat edges. The assessment was

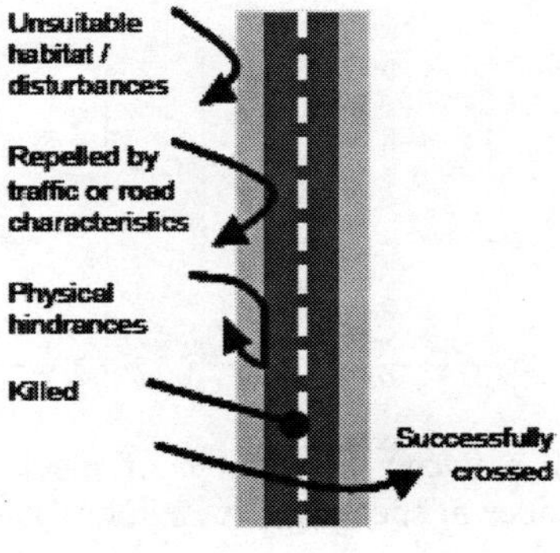

Figure 2: The barrier effect of a road. Source: [8].

restricted to fauna. The first impact is the **barrier effect.** Traffic imposes dispersal and migration range barriers to most non-flying terrestrial animals and causes death while crossing the road [8]. The barrier effect is a combination of disturbance and avoidance effects, physical hindrances and traffic mortality, together reducing the number of movements across the barrier (Figure 2).

The barrier effect is a non-linear function of [10]:

- traffic intensity: from 6.000-8.342 per 24 hours in the current situation to 17.200 in case the highway would be built;
- average vehicle speed: from 70 to 90 km/h;
- road width: from 9 to at least 25 m;
- roadside characteristics: more hard elements;
- type of species, behavior, sensitivity to disturbances.

When the traffic intensity would increase if the current road would be replaced by the highway, the road would become an absolute barrier to cross for some animals, as Table 1 shows. In that case the animal species cannot cross the barrier and populations on each side of the road become isolated. When, as a consequence, habitats on each side of the rode become too small (see Figure 3) the specie will become extinct in the area.

Many mammals will not be able to pass the road. Birds are also less likely to fly over the road. We assume an increase of the barrier effect by a few percentages.

Table 1: Barrier-effect of roads: traffic intensity values where roads become absolute barriers to cross.

Species	*Traffic intensity (vehicles/day)*
Lizard, viper, reddish vole	2.000
Northern vole, squirrel	3.500
Marten, badger, roe-deer, fox	12.000
Red deer, swine, otter	15.000

Source: [9].

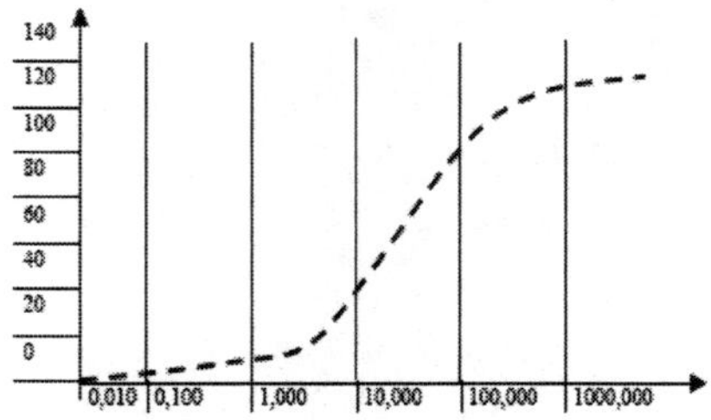

Figure 3: The relation between the number of species and the surface of an area (y: number of species, x: area in km^2). Source: [9].

Disturbance. When the average vehicle speed increases by 10 km/h and the traffic intensity increases as expected, the area affected increases by 44 m [11]. If we take the change in bird densities as an indication for the increased disturbance of the ecosystem, then the disturbed area will increase by 74.8 ha.

Mortality effect. The number of road kills generally increases with traffic intensity. High mortality will occur among amphibians and mammals, especially bear and wolf. Next to this insects, birds and bats are also affected [8]. We expect an increase by a few percentages.

Habitat loss. The net loss of wildlife habitat would be at least 27,2 ha ([25 – 9] x 17 km). If the highway is constructed in the narrow bed of the gorge, most of the current natural habitats along the Struma River will be destroyed.

Artificial lightning. This effect is not easy to quantify, but we assume the spatial burden to be limited to the direct surroundings of the highway.

Depending on the number of successful crossings relative to the size of the population, the barrier effect may affect the populations of species. If the exchange of individuals is further reduced but not completely inhibited, the populations may diverge in characteristics such as density, sex ratio, birth and mortality rate. Also genetic differences may emerge, as the chance for mating with individuals from the other side of the road barrier is reduced. Fragmentation may lead to inbreeding witness studies on rodents and amphibians. A barrier becomes absolute for a species when crossings stop. If isolated areas become too small to live for a certain species, then it becomes extinct in the region [9].

Table 2: Ecological impacts of the highway.

Effects	**Variable (# of)**	**↑/↓**
Barrier effect	Unsuccessful crossings	± 5% ↑
Disturbance effect	Ecologically disturbed area	74.8 ha ↑
Mortality effect	Kills per unit of time	± 5% ↑
Habitat loss	Destroyed habitat	> 27.2 ha ↑
Artificial lightning	Disorientation, fixation	↑

While many species will disappear, some may benefit. For instance, road verges can be beneficial to animals and plants, depending on the type of roads. Quiet roads with little traffic are expected to be more beneficial than highways.

3.4.2 Environmental (a-biotic) impact

The first effect concerns **disruption of natural processes.** Quantification of this effect is difficult, however.

Air contamination. Examination of sensitive organisms demonstrates that the current road affects the slopes along the Struma River up to 2 km away. Higher traffic intensities may be balanced by less congestion. We assume a certain increase in contamination due to the highway.

Adverse effects from construction, maintenance and use of the road. The size of the affected area is likely to increase.

3.4.3 Impact on size and nature of recreation

Several (eco-)activities are under development. A visually less attractive and noisy landscape will attract less visitors. A relatively comparable case is the reactivation of the Dutch section of the Iron Rhine railway [12]. From it we estimate the number of recreational visits to fall by more than 10 per cent.

Table 3: Environmental impacts of the highway.

Effects	Variable (# of)	↑/↓
Disruption of natural processes (ground water etc.)		↑
Air pollution	Mg/ltr air	↑
Adverse effects from construction, maintenance, use of the road		↑

Table 4: Recreational impacts of the highway.

Effects	Variable (# of)	↑/↓
Less attractive recreational environment	Recreational visits	Region: ↓10% [1)]
		Nation: ± 0

Note: 1) Very indicative. Eco-tourism will be affected, because the Struma river is a favorite area, as is the large Melo Sand Hill near Kresna town.

Table 5: Socio-economic impact.

Effects	Variable (# of)	↑/↓
Income and wealth	Decrease of agricultural land, accessibility	0/↓ [1)]
Human health	Pollution of air, soil, water	↑
	Risk for drivers and trespassers	↑ [2)]
	Noise disturbance to humans	↑ [3)]

Notes:

1) 0 if the road is built in the nature area only, else a decrease. Junctions would improve accessibility, but they are not planned.
2) No secondary road or (level) crossings are planned. Tractors and horse drawn carts are mainly used by farmers. Accidents with larger animals at high speed will be more frequent and more dangerous.
3) The road is very close to the town of Kresna. The value of houses will be reduced.

3.4.4 Socio-economic effects

To be mentioned are the following impacts:

Agriculture. We assume that the highway will be built mainly in the nature area.

Health effects. Exposure to aerosols is accounted for earlier deaths of thousands of people each year [13].

Risk for drivers. Police records in Europe (excl. Russia) suggest at least 0.5 m. ungulate-vehicle collisions per year, or at least 300 human fatalities, 30.000 injuries, and material damage of more than US$ 1 bn. [8] The gorge inhabits large mammals like bear, wolf and deer.

Noise disturbance. Long term exposure to noise can induce psychological stress and eventually lead to physiological disorder.

3.5 Economic evaluation

Ecological and environmental impacts are not revealed by market prices as they concern nonuse values. We used a set of methods best fitting to the purpose.

Barrier effect. Method: avoidance cost. We assume two ecoducts of 60m length, 30m width and 5m height with a cost of 2 x 1.7 million [14].

Disturbance effect. Method: avoidance cost. Noise shields cost approximately € 60.000 for a shield of 3 m high and 150 m long [15].When applied to the 17 km E-79 the cost would be 2*(17.000/150)* €60.000 = € 13.6 m.

Habitat loss. Method: Restoration cost. € 272.000.

Artificial lightning. Method: avoidance cost. Adapting artificial lightning (by limiting direction and amount of light, adaptation to traffic intensity, etc.). Investment cost of approximately € 80,000 per km of road [15]. When applied to the 17 km E-79 the total costs would be 80.000*17 = € 1.36 m.

Environmental effects. The appropriate method would be CVM. No data were available, hence we used a pro memorie (PM) as proxy.

Impact on recreation (use values). Methods: Market valuation, travel cost. Impact on recreational spending will be negligible if we assume substitution on a national scale. The welfare loss is due to the decreased recreational value of the sites, hence shift of travel to other sites. No data is available, hence a PM.

Socio-economic effects. We assume a negligible impact on agriculture. Human health will be affected. Air pollution could best be valued by hedonic pricing, as no data is available we apply PM. Noise disturbance and risks for drivers are included in the disturbance of the ecosystem.

Total cost: € 3.4 m. + € 13.6 m. + € 0.272 m. + € 1.36 m. + PM = > **€ 18.7 m.**

3.6 An alternative

An alternative route via the Pirin Mountain could prevent all the negative effects. Its impact will be much less than the existing road, which will then be converted into a disclosure route for inhabitants and tourists [5].

EU's environmental regulators can play a vital role in protecting this valuable area against strong economic interests [6].

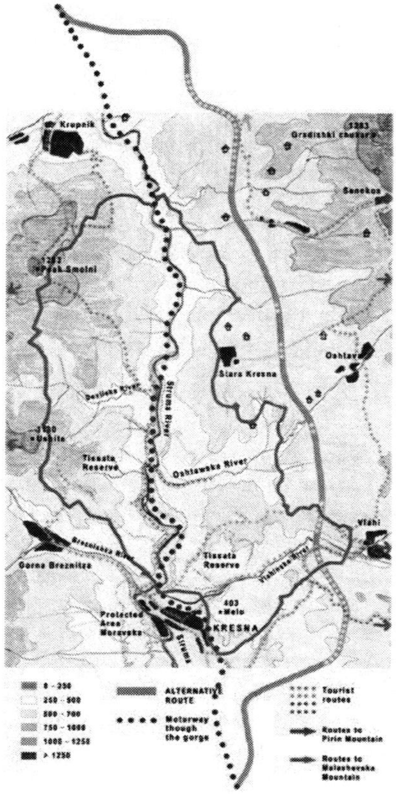

Figure 4: Map of the Kresna Gorge with eastern bypass. Source: [4].

4 Evaluation

In this paper we presented the results of a study into the impact of road extension on a rather unique and irreplaceable nature area. The importance of this paper lies in the depth of the ecological analysis and valuation on the one hand and the combination of several valuation methods on the other hand. CBA studies tend to concentrate on issues like noise, air pollution, use of space and socio-economic impact. This study went much further. By combining biotic, a-biotic and socio-economic impacts it was possible to give a relatively complete overview of the impacts of the highway. This way of analyzing gives clues to interesting areas for further research, in particular into ecological and environmental impacts of infrastructure.

The valuation of the external costs of the highway shows a relatively small value of € 18.7 m. This is due to a number of factors:
- valuation problems, which lead to an incomplete dataset (PM);
- the prime use of the cost avoidance method, which is a valuation from a human perspective.

The impact of the existing road is already considerable. The impact of the highway is therefore attenuated. A more cost-intensive in depth study should be undertaken in the gorge in order to remove the PM's from the cost calculation.

5 Conclusions and recommendations

A few methodological observations can be made. First, infrastructure projects in nature areas particularly affect nonuse values. Second, avoidance and restoration costs methods are most useful when valuing adverse ecological effects in economic terms. These methods are not based on consumer preferences (wtp), but on observed market prices. These methods can be used to value effects that have been mitigated elsewhere.

An issue for further research is to focus on effects where avoidance cost based data are not available. For such effects contingent valuation could be applied in order to get an indication of the size of these effects in reference to other costs and benefits. A recommendation for policy making is to include effects on surrounding nature areas into the CBA for infrastructure. From our study it is clear that such effects can no longer be omitted from an integral evaluation.

By building infrastructure in nature areas, we threaten the existence of very unique, complex and irreplaceable ecosystems. Unlike built-up areas, which can 'bounch back' after a road is ready, such ecosystems do not recover after the road is finished. The best advice would therefore be to stay out of such areas wherever possible. This is what nature protection is meant to do. To improve the present situation while taking care of the needs of increased (international) traffic and the economy, this highway should be built elsewhere.

References

[1] Vleugel, J.M., and E.J. Bos, 2008, Ways to deal with the 'temporary value of cost benefit analyses, in: K. Aravossis, C.A. Brebbia and N. Gomez (eds.), Environmental Economics and Investment Assessment II, Wessex Institute of Technology, Ashurst, UK, pp. 171-180.

[2] Nikolov, S. and S. Spasov, 2005, Frequency, density and numbers of some breeding birds in the south part of Kresna Gorge (SW Bulgaria), **Acrocephalus** 26 (124), pp. 23–31.

[3] http://www.kresna.org/index_en.php.;~/motorway_en.php; ~/gallery_en.php, ~/alternative_en.php.

[4] Bos. E.J, 2008, Valuation of Ecological Networks - Case study Kresna Gorge Bulgaria, Alterra and LEI Working Report 1896, Wageningen.

[5] http://www.kresna.org/index_en.php and inked pages on the website.

[6] EU DG TREN, 2008, TEN – T Trans-European Transport Network, Implementation of Priority Projects Progress Report, May 2008, Brussels.

[7] http://en.wikipedia.org/wiki/Struma_motorway.

[8] Seiler, A., 2001, Ecological Effect of Roads: a Review. SLU, Uppsala.

[9] Pouwels, R., R. Jochem, M.J.S.M. Reijnen, S.R. Hensen and J.G.M. van der Greft, 2002, LARCH voor ruimtelijk ecologische beoordelingen van landschappen, Alterra, rapport 492, Wageningen.
[10] Berthoud, G., 2002, Construction of a Motorway in the Gorge of Kresna Corridor Connection Bulgaria – Greece: Motorway E79: Sofia-Kulata. Report of the on-the-spot appraisal. Standing Committee 22nd meeting Strasbourg, 2-5 December 2002.
[11] Reijnen, M.J.S.M., G. Veenbaas and R. Foppen, 1992, Het voorspellen van het effect van snelverkeer op broedvogelpopulaties. DWW-rapport, Rijkswaterstaat Delft/IBN-DLO, Wageningen.
[12] Konijnenburg, P. van, J. Kortman, J. Jantzen and H. van der Woerd (2001). Effecten van Reactivering van de IJzeren Rijn: Onderzoek naar de Effecten van Reactivering van de IJzeren Rijn op de Functies Recreëren, Wonen en Werken, IVAM, Amsterdam.
[13] Slanina, S. and W. Davis, 2008, "Impact of local air pollution.", in: Encyclopedia of Earth, C.J. Cleveland (ed.), Environmental Information Coalition, National Council for Science and the Environment, Washington, D.C.
[14] Vanya Simenova, personal information from Bulgaria.
[15] Rijkswaterstaat, Dienst Weg- en Waterbouwkunde, 2002. Effecten en kosten van leefbaarheidsmaatregelen, Den Haag.

Implementation of the polluter pays principle – example of planning for decommissioning

S. Lindskog[1] & R. Sjöblom[2]
[1]*The Swedish Radiation Safety Authority, Sweden*
[2]*Tekedo AB, Sweden*

Abstract

The Swedish Radiation Protection Authority (SSM) and some of its predecessors have since the late nineteen seventies overseen the Swedish system of finance for decommissioning and waste management of nuclear facilities. This system contains segregated funds for the costs according to best estimate and securities to cover uncertainty.

Recently, the underlying legislation was extended to also include various small facilities with sometimes small businesses as owners, and the Government authorized the SSM to issue regulations as warranted and appropriate.

The implementation of the new legislation includes the challenges of simultaneously honouring the polluter pays principle and the principle of equity between the generations whilst at the same time complying with the requirements on proportionality as well as harmony with other legislation.

Surveys have therefore been conducted regarding similar solutions in other areas as well as statements in other legislations, and the results are briefly summarized in the present paper. Previous supporting work includes analyses of planning for decommissioning and cost calculation methodologies.

It is found that the estimated cost, prepared in accordance with the state of the art, can form the basis for the selection of means for financial assurance. Thus, exemption can be recommended for liabilities up to k€ 2,4, securities alone up to M€ 0,1, and securities in combination with segregated funds above this level.

It is commented that some ombudsman type of organisation is required to safeguard the interests of future generations with regard to environmental liabilities, and that advice may be received from the younger generation.
Keywords: decommissioning, nuclear, fund, security, Sweden, legislation, polluter pays principle.

WIT Transactions on Ecology and the Environment, Vol 131, © 2010 WIT Press
www.witpress.com, ISSN 1743-3541 (on-line)
doi:10.2495/EEIA100031

1 Definitions

There are three levels of legislation in Sweden:

Legislation	Issued by	Compliance with	Authorized by
Law	Parliament	Constitution	Swedish people
Ordinance	Government	Laws	Parliament
Regulation	Competent Authority*	Laws and Ordinances	Government

* Such as the Swedish Radiation Safety Authority.

Laws, ordinances and regulations are legally binding and the compliance of them is overseen and assured by our legal system, including our courts.

In addition, a Competent Authority can issue general advice with regard to a certain regulation. It can contain clarification as to what the actual regulation is intended to mean and may also provide examples. General advice is not legally binding and compliance must not necessarily be upheld in a court decision.

Competent Authorities – like everybody else, e.g. a branch organization – can also issue guidance documents. They reflect good practice, but cannot necessarily be relied on for compliance with legislation.

In this paper, the Swedish Radiation Safety Authority (in Swedish: Strålsäkerhetsmyndigheten) is referred to by its abbreviated name, SSM. This paper refers to work in progress. Any conclusions represent the views of the authors.

2 Introduction

2.1 Implementation of environmental legislation

The last few decades have meant a shift of paradigm in that basic principles have been established and policies agreed on a number of environmental issues including protection of health and environment, conservation, re-cycling, sustainable development, remediation, use of best available technology, equity between generations and the polluter pays principle (PPP), also known as the Extended Polluter Responsibility (EPR).

Associated legislation has been issued and enforced. Releases to the environment – e.g. as fly ash and sulphur dioxide – have plummeted.

The success rate is very varied, however, and the following was written [1] in 1997 by Staffan Westerlund, professor of environmental law at the University of Uppsala (translation from Swedish by the present authors):

"It is well known that environmental laws seldom function well and that environmental goals are usually not achieved. We have also become accustomed to an almost total inefficiency of regulations intended to alter environmentally inappropriate behaviour. It does not come as a surprise that concrete rules ... still 25 years after having come into force have not been implemented and enforced. Over the years, there have been so many incidences of malfunctioning of the environmental legal system that we who teach law must make a quite clear

distinction between on one hand the law as it is written,... and on the other hand how it actually functions (or actually does not function)."

Obviously, the prerequisites for implementation of general strategies and framework laws can vary enormously between different situations.

Cases are comparatively straightforward and easy where

- The potential polluter and its activities can be clearly identified;
- The impact on health and environment can readily be measured and assessed;
- The potential polluter needs a permit (with certain conditions);
- The potential polluter is competent and rational in the selection of the best available technology.

Cases are comparatively difficult, and even cumbersome, when the link between the polluter and the potential impact is difficult to establish, e.g. when the relations between cause and effect are not well understood. Another difficulty arises when remediation is to be done long after the corresponding industrial activity has been terminated. Implementation may be further impeded if, in addition, technical and thereby also financial planning is difficult to carry out long before the restoration is to take place.

Consequently, cases where the polluter pays principle is to be implemented simultaneously with the principle of solidarity between generations can be expected to be particularly daunting, especially if there are also obstacles to the planning process.

2.2 Nuclear legacy legislation and its implementation

All these challenges apply when these two principles are to be implemented for the case of decommissioning of old nuclear research facilities (and similar) in Sweden.

Previous legislation on financing of our nuclear legacy has included our ten nuclear power plants and certain major facilities from our nuclear technology development period (about 1955 to 1975). The new legislation, primarily in the form of an ordinance [2] that came into force on November 1st 2008, includes all nuclear facilities regardless of size. The ordinance also contains authorization from the Government to the SSM to issue regulation as warranted and appropriate for the implementation.

In addition to proper authorisation, our Constitution states a number of requirements that apply to any such regulation, including the following:

- A regulation must contain a reasonable balance between different interests, and the benefits must be justified in comparison with the costs for compliance. (The proportionality principle)
- All must be dealt with in an equal manner.
- There must not be any contradictions with any other legislation.
- There has to be a follow-up of the outcome, and adjustments made as appropriate from any lessons learned.
- A regulation must be simple and clear.

The first bullet in this list corresponds to the so-called proportionality principle which must be applied simultaneously with the other principles.

The present work was prompted as a result of the above mentioned authorization from the Government.

3 Purpose and scope

The purpose of the present paper is to illustrate how the polluter pays principle and the principle of solidarity between generations can be implemented simultaneously even in cases where the planning is difficult.

The scope of this paper is to summarize the following

- the implementation of these principles in several areas
- relevant legislation
- prerequisites for financial planning for decommissioning of nuclear facilities

The scope is also to analyse this information for the purpose of implementation of the two principles. This will be carried out with small nuclear facilities as an example.

4 Environmental liabilities in some areas

4.1 Forestry

Legislation on preservation and protection of forest extends beyond written records, and statements to this end exist even in our oldest written law, *the older Westgothia law* [3], from around the year 1220.

Early environmental legislation includes the statement by our queen Kristina on March 18th 1639 when she banned burn-beating by the penalty of banishment. The wood from the forest had better use in our metals beneficiation industry.

The Swedish Forestry Act [4] was first issued in 1903, at around which time our forests were being converted from primeval (old-growth) forests to forests that are actively managed (including recurrent replanting). The law states that the owner must replant after harvest. When harvested forest is sold, the seller may have to guarantee that the buyer will replant by means of securities.

4.2 Mining, beneficiation and contaminated soil

Our first record of a mining stock company relates to the Falun copper mine and is from the year 1288 [5, 6]. The roasting of this sulphide ore gave rise to enormous emissions of sulphur dioxide, as has been witnessed by many travellers including Carl von Linné (who developed a system for the categorization of plants). The smoke was bothersome to the miners and was disastrous to the crops around, but had the advantage that plague never reached the area. At least during most of the 17th and 18th century there was a constant legal battle between the miners and the farmers on the emissions, and for long periods of time ore could be roasted only outside the growing season.

There are many old mine tailings and associated contaminated soil in Sweden that need remediation. Contaminated soil has also come about as a result of many other activities. At present, annual public spending in Sweden on remediation of old mine tailings and contaminated soil exceeds M€ 50. According to the website of our Environmental Protection Agency (in Swedish: Naturvårdsverket) this is insufficient even to remediate those sites associated with the highest risks before the year 2050.

For present day industrial activities, the legal entities associated with mining and beneficiation activities are fully responsible for any remediation required. Copper ore is no longer mined at the Falun site, but at Boliden and Aitik. According to the Annual Report for the year 2008 of Boliden AB (AB is an abbreviation of the Swedish word "Aktiebolag" which means limited stock company), the Competent Authority has received guarantees for around M€ 50 to cover the environmental liabilities for the restoration associated with the Aitik mine alone.

Especially residues from mining of sulphide ores have the potential of causing increased releases over time as a result of the acidification that takes place if and when the residues come in contact with the oxygen in the air.

The heap of shale ash at Kvarntorp near Örebro constitutes an example of such a situation. It is a hundred metres high and comprises several tens of million tonnes of material. The ash was generated during 1940 and 1965 when oil was extracted from the shale. Much of the carbon and sulphur in the shale was left un-combusted in the processes. The residues are burning still today, and for this reason, the heap is kept dry and the releases are low. If no remedial action is taken, the heap will cool sooner or later, and it can then be feared that acidification takes place and the releases become substantially increased.

At the same time, it should be realized that there is a substantial difference between on one hand the cautious modelling made in order to ensure that environmental impact is acceptable, and on the other hand realistic modelling. The streams and lakes surrounding the Falun copper mine mentioned above have a surprisingly healthy biological life [5].

4.3 Offshore

There are no offshore rigs for oil production in the Swedish waters, and consequently no associated domestic need for decommissioning and restoration.

Internationally, offshore is one of the major areas of environmental liabilities. It is no longer acceptable to just tow a scrap rig to deeper waters and sink it. Instead, all structures above the bottom of the sea have to be taken ashore and either be recycled or deposited in a landfill.

The decommissioning is organized somewhat differently as compared to other areas. Oil rigs represent enormous investments whilst the return in terms of oil production may be variable. Consequently, even large oil companies often own and operate oil rigs through consortia. In concordance, planning for decommissioning as well as issuing of financial assurances to Competent Authorities are also carried out through consortia based on consortia agreements.

Offshore decommissioning includes the challenges of deep water work with aged and possibly deteriorated large structures. However, most of the material retrieved is recycled, and residues are deposited mainly in ordinary landfills. Thus, offshore decommissioning does not have to deal with such sometimes large uncertainties that are associated with the decommissioning of nuclear facilities and their associated radioactivity, nor does it have to take responsibility for long-term events in waste repositories such as may be the case for mining residues, contaminated soil and nuclear waste.

Much of the information presented above on offshore decommissioning has been obtained through a membership of one of us with the Norwegian Petroleum Society (In Norwegian: Norsk Petroleumforening), see www.npf.no.

4.4 Nuclear technology

Sweden was one of the six countries that took part in the rush to build the first nuclear power stations. The other countries were United States, United Kingdom, France, Soviet Union and Canada. All other countries with nuclear power reactors have had to turn to either of these for assistance [7]. Our first nuclear power plant was taken into operation in 1963 as a result of a comprehensive research and development (R&D) programme. The contributions to this programme from the Government alone (during the years 1955–1975) amounted to a total of around G€ 1,55 in today's currency [8]. Most of the R&D work was conducted in the facilities at Studsvik, see Figure 1 [9–13].

The first controlled chain reaction took place in 1942, almost 50 years after the x-rays (= gamma rays) had been discovered. The hazard associated with radiation (e.g. cancers many years after exposure) was well known at that time, but there was no experience with induced radioactivity. It took until the mid seventies until nuclear waste research caught on speed. According to the national implementer of this R&D programme, the Swedish Nuclear Fuel and Waste Management Company, SKB, the accumulated R&D cost until present at the price level of today is about G€ 1,90 (see www.skb.se). This research covers the waste from our ten nuclear power plants in operation and three taken out of service as well as the waste from research and development.

Old literature, see e.g. "*the architect in the nuclear age*" from 1964 makes no or little mentioning of the implications of induced radioactivity for decommissioning of a facility [9]. Actually, decommissioning may be even more expensive than building a facility, especially if no precautions were taken initially.

The total cost for decommissioning and waste management of the facilities in Figure 1 is presently estimated to around 0,2 G€. This includes the cost for decommissioning of the Active Central Laboratories which might be estimated to around M€ 10 (no final report is known to the authors). The estimates on the costs remaining in 1998 have varied up to a factor around three, largely due to difficulties in assessing the varying alpha contamination in the absence of good fingerprint gamma emitters [14–16].

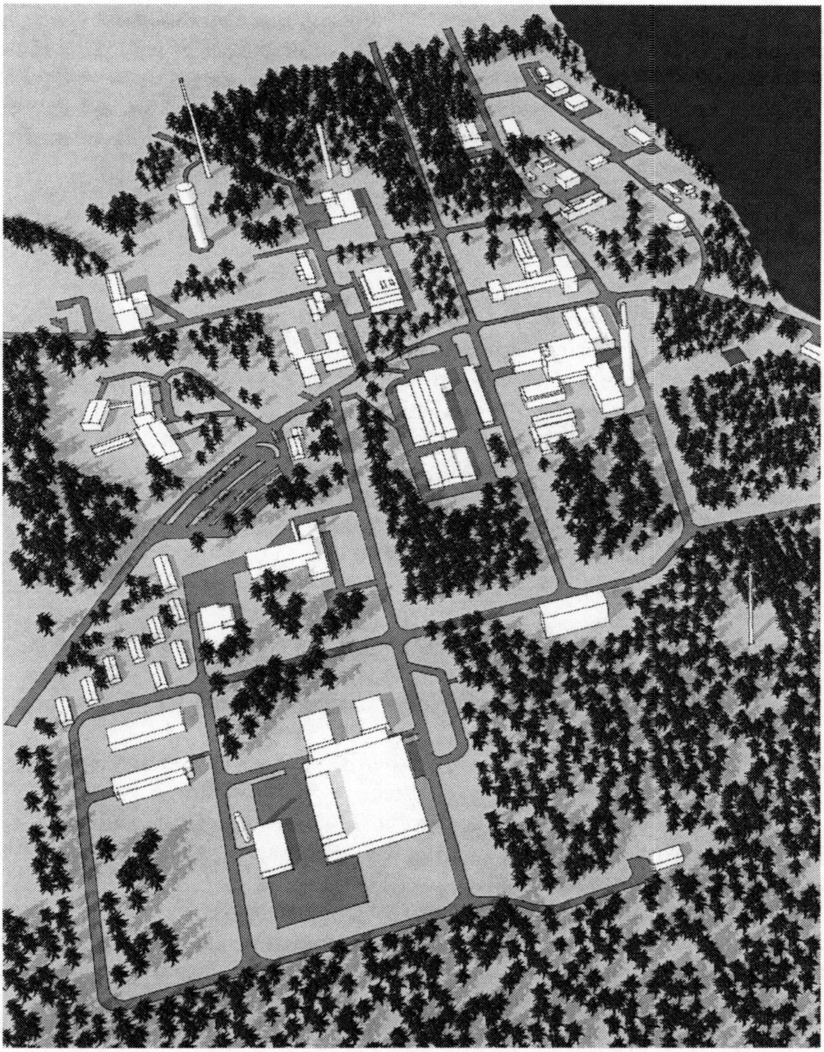

Figure 1: Artist's impression of the Studsvik site (about 80 kilometres south of Stockholm) at around the year 1965. The illustration is based on contemporary drawings [9, 10] and areal photographs [10–13]. (The site looks substantially different today.) The size of the site is approximately 1,0 by 0,7 km^2. The lower middle block contains the two buildings that comprised the Active Central Laboratories (ACL). It has recently been decommissioned to green field conditions, cf text. The laboratory was built for mixed oxide (uranium + plutonium) fuel fabrication on a laboratory and a pilot scale, but equipment for the latter was never installed.

The total nuclear legacy in Sweden is at present estimated to about G€ #. The uncertainty (in %) is much larger for small and old facilities as compared to new and large ones. The costs are to be covered by two segregated funds, one for the 12 modern nuclear power plants (NPPs) and one for most of the old facilities from the era of nuclear technology development. The funds also have securities to cover uncertainty.

5 Legislation

5.1 Nuclear technology legislation

Nuclear safety is primarily regulated in Sweden by and under two laws: The Radiation Protection Act [17] and Act on Nuclear Activities[18]. The Radiation Protection Act applies generally to radiation.

The Act on Nuclear Activities is valid for facilities in which nuclear chain reactions take place and related facilities. It is also valid for nuclear material (i.e. material that is fissile or can be activated to become fissile), for activated material (with several exceptions), and for nuclear waste.

The Nuclear Liability Act [19] is valid for nuclear material that is not intended to be reused and for nuclear waste that is not waste from daily operation.

As already mentioned, there are two "compartments" for securities and fees to segregated funds:

- A the anticipated costs for decommissioning and waste management etc, and
- B a risk fee intended to cover the risk that the Government takes in its management of the fund system.

Compartment A comprises a combination of securities and assets in segregated funds. Securities are lifted at the same pace as that of the payments that flow into the segregated funds.

The new ordinance [2], issued by the Government under the Nuclear Liability Act [19], states that those eligible under the law who are not nuclear power reactor owners are obligated to submit to SSM every third year a cost calculation comprising the following:

- the total best estimate for the cost for decommissioning and waste management,
- the expected remaining time of operation,
- the proposed proportions between securities and assets in a segregated fund.

Subsequently, the SSM will review the material and decide on the fee to be paid.

The new ordinance has also granted the SSM authority to decide on exemption from requirements on securities as well as on payments to segregated funds.

5.2 The Swedish environmental code

The Swedish Environmental Code states [20] as follows: "*Persons who pursue or have pursued an activity or taken a measure that causes damage or detriment to the environment shall be responsible, until such time as the damage or detriment ceases, for remedying it to the extent deemed reasonable ...*". The Code also states that permits issued under the code may be associated with requirements on securities corresponding to all future costs.

5.3 Financial reporting legislation

The laws covering financial reporting for ordinary companies in Sweden are primarily the Accounting Act [21], Annual Reports Act [22] and the Swedish Companies Act [23]. The Accounting Act states in §2 that the obligation of book-keeping must be carried out in accordance with "good practice".

Large companies are obligated to follow the International Financial Reporting Standards and International Accounting Standards (IFRS/IAS) [24] while small companies may follow the general advice issued by Swedish Accounting Standards Board (in Swedish: Bokföringsnämnden, BFN) [25].

5.4 Criminal law

The Swedish Penal Code[26] is the same for all. It states that anyone who is obligated to follow the Accounting Act [21] but declares figures that are not correct so that the books no longer present an "essentially correct financial situation" may be sentenced to jail for at most six years (in severe cases).

There is a clear possibility that errors and uncertainties in estimations of decommissioning costs may be large in comparison with the errors and uncertainties for other posts in financial books. Compliance might be evaluated using the Elofsson method, according to which an acceptable deviance may be at most 30% [27].

6 Example of small nuclear facilities

It was mentioned in Section 2.2 that the new ordinance [2] authorizes the SSM to issue regulation as warranted and appropriate for its implementation. The ordinance also refers to present and previous holders of permits under the Act of Nuclear Activities [17], but leaves it to the SSM to regulate what tools to apply and to differentiate between them.

It might be tempting to assume that such selections should be based on the complexity of the nuclear activity that has taken place in any particular facility. Indeed, it can be concluded that the decommissioning of a nuclear facility would be quite trivial if only sealed sources have been handled or if all radioactivity has been short-lived. Such cases could warrant exemption. Otherwise the relation has been found to be weak, however.

Instead, it has been found that differentiation between funds, securities and exemption can be based on estimations of the cost for decommissioning. As has been briefly illustrated above, such costs have proven notoriously difficult to estimate with any accuracy, especially for old research facilities. It has been found, however, that a precision of ±15% might be attained, at least in favourable cases, if adequate planning is carried out. The prerequisites for this were presented at the previous meeting on Environmental Economics [28], see also [16, 29–33] and references therein.

Such planning is required already under the financial laws (cf Section 5.3) and the penal law (cf Section 5.4). To use estimated costs as a basis for selection between funds, securities and exemption does thus not impose any new or further burden on the facility owners. It is therefore the alternative put forward in the present study. It is also suggested that the upper limit for exemption be set at around k€ 2,4 and the upper limit for securities alone at around k€ 100.

The Nuclear Liability Act [19] as well as the new ordinance [2] regulate how financing of decommissioning is to be carried out in general and specifies that segregated funds supplemented by securities are to be used.

7 Final comments

Application of the polluter pays principle simultaneously with the principle of equity between generations requires considerably more complex considerations than e.g. a release limit in that adequate attention need to be applied in several areas simultaneously. In the example above of decommissioning of small nuclear facilities, the timing of the planning is largely dictated by the needs of the system of finance.

It might be tempting to consider that such a sustainable approach is a modern innovation. However, according to the Westgothia law [3], land belonged to the clan and was to be passed over to the sons and daughters when they married. It could be sold outside the clan only after consent from the relatives together with a court ruling. Instead, the quarterly report syndrome - where company managements follow the whims of institutional investors rather than the long-term prerequisites – seems to be a modern phenomenon.

Research suggests that an individual will sacrifice consumption to benefit future generations only if the guarantee exists that others will also do so [34].Thus, modern bodies are needed for the ancient tasks of the clan and the court. Such solutions are proposed in Reference [34].

It is important in this regard to realize that we do have access to the values of one future generation, namely the young generation, see a separate paper to the present conference[35].

References

[1] Westerlind, S., *A legal system for sustainable development.* (In Swedish: En hållbar rättsordning). Justus förlag, Uppsala 1997. ISBN 91-7678-372-3.

[2] *Ordinance on financial action for the management of residues from nuclear technology activities.* (Förordning om finansiella åtgärder för hanteringen av restprodukter från kärnteknisk verksamhet, in Swedish). SFS 2008:715.

[3] *The older Westgothia law*. (In Swedish: Äldre Västgötalagen, översatt och förklarad av Nat. Beckman). Appelbergs Boktryckeri Aktiebolag, Uppsala, 1924.
[4] *The Swedish Forestry Act*. (In Swedish Skogsvårdslagen) SFS 1979:429.
[5] Lindeström, L. *The environmental history of the Falun Mine*. Almqvist & Wiksell Tryckeri, Uppsala 2003. ISBN 91-631-3536-1.
[6] Lindroth, S., *Mining and copper beneficiation at the Great Copper Mountain*. (In Swedish: Gruvbrytning och kopparhantering vid Stora Kopparberget). Almqvist & Wiksell Boktryckeri AB, Uppsala 1955.
[7] *Controlled nuclear chain reaction, the first 50 years*. American Nuclear Society, 1992. ISBN 0-89448-557-1.
[8] Fjæstad, M., & Jonter, T., *The Rise of the Nuclear System of Innovation in Sweden*. Svenska ekonomisk-historiska mötet i Stockholm 2007.
[9] Munche, J-F., *The architect in the nuclear age. Design of buildings to house radioactivity*. Iliffe Books Ltd., London, 1964.
[10] *Studsvik, Swedish research establishment*. EuroNuclear, September, 1965.
[11] Österlundh, C. G. & Erwall L. G., *Erzeugung und Anwendung von Radioisotopen*. (In German; Production and use of radioisotopes) Die Atomwirtschaft, February, 1963.
[12] Aler, B., *Survey of Sweden*. Nuclear Engineering International, September 1970.
[13] Bladh, R., and Eriksson, O., *40 år I Studsvik*. (In Swedish; 40 years at Studsvik). Studsvik AB. ISBN 91-7010-283-X.
[14] Jonsson, B., Bergström, L. & Lindberg, M., *Decommissioning of the ACL and ACF plants in Studsvik, Sweden*. Waste Management '04 Conference, February 29 – March 4, 2004, Tucson, Arizona, USA.
[15] Hedvall, H. R., Stridsman, K. H., Berg, R. S. & Johnsson, B., *Project evaluation of the decommissioning of a laboratory plant at Studsvik*. Waste Management '06 Conference, February 26 – March 2, 2006, Tucson, Arizona, USA.
[16] Lindskog, S. & Sjöblom, R., *Radiological, technical and financial planning for decommissioning of small nuclear facilities in Sweden*. Proceedings of the 12th International Conference on Environmental Remediation and Radioactive Waste Management, ICEM2009, October 11-15, 2009, Liverpool, UK.
[17] *Act on Nuclear Activities*. (In Swedish: Lag om kärnteknisk verksamhet). SFS 1984:3.
[18] *Radiation Protection Act*. (In Swedish: Strålskyddslag). SFS 1988:220.
[19] *Nuclear Liability Act*. (In Swedish: Lag om finansiella åtgärder för hanteringen av restprodukter från kärnteknisk verksamhet). SFS 2006:647.
[20] *The Swedish Environmental Code*. English translation. Ds 2000:61. (In Swedish Miljöbalk, SFS 1998:808).
[21] *Accounting Act*. (In Swedish: Bokföringslag). SFS 1999:1078.
[22] *Annual Reports Act*. (In Swedish: Årsredovisningslag). SFS 1995:1554
[23] *The Swedish Companies Act*. (In Swedish: Aktiebolagslagen). SFS 2005:551.

[24] *International Financial Reporting Standards and International Accounting Standards* (IFRS/IAS). International Accounting Standards Board. 2008.
[25] *Bokföringsnämndens allmänna råd om årsredovisning i mindre aktiebolag.* (General advice on annual reporting in small companies issued by the Swedish Accounting Standards Board, In Swedish).
[26] *The Swedish Penal Code.* (In Swedish: Brottsbalk). SFS 1962:700.
[27] Dahlqvist, A-L. & Elofsson, S., *Crimes in accountance and the law.* (Swedish title: "Bokföringsbrott och bokföringslagen"). Norstedts juridik, Stockholm, 2005. ISBN 91-39-10709-4.
[28] Lindskog, S., Cato, A. & Sjöblom, R., *Estimations of costs for dismantling, decommissioning and associated waste management of nuclear facilities, and associated impact on decision processes, functioning of markets and the distribution of responsibilities between generations.* Environmental Economics II, 28-30 May 2008, Cadiz, Spain. WIT Transactions on Ecology and the Environment, Vol 108. Wit Press, 2008. ISBN 978-1-84564-112-2.
[29] Lindskog, S. & Sjöblom, R., *Regulation evolution in Sweden with emphasis on financial aspects of decommissioning.* Decommissioning Challenges: an Industrial Reality? Sept. 28 to Oct.2, 2008 – Avignon, France.
[30] Iversen, K., Salmenhaara, S., Backe, S., Cato, A., Lindskog, S., Callander, C., Efraimsson, H., Andersson, I. & R. Sjöblom, R., Cost calculations at early stages of nuclear facilities in the Nordic Countries. The 11th International Conference on Environmental Remediation and Radioactive Waste Management. September 2-6, Bruges (Brugge), Belgium.
[31] Sjöblom, R., Sjöö, C., Lindskog, S. & Cato, A., *Early stage cost calculations for determination and decommissioning of nuclear research facilities.* The 10th International Conference on Environmental Remediation and Radioactive Waste Management. Glasgow, UK, 4-8 September, 2005.
[32] Cato, A., Lindskog, S. & Sjöblom, R., *Financial Planning as a Tool for Efficient and Timely Decommissioning of Nuclear Research Facilities.* American Nuclear Society. Decommissioning, Decontamination and Reutilization. Capturing Decommissioning Lessons Learned. September 16-19, Chattanooga, Tennessee, USA.
[33] Laraia, M. & McIntyre, P. J., responsible officers; Cato A., Lindskog S. & Sjöblom. R. et al contributors. *Decommissioning of research reactors and other small facilitiesby making optimal use of available resources.* IAEA Report Series 463, Vienna 2008.
[34] Padilla, E., *Intergenerational equity and sustainability.* Ecological Economics 41 (2002) 69-83.
[35] Labor, B. & Lindskog, S., *Values held by young stakeholders on financial planning regarding liabilities for nuclear decommissioning.* Environmental Economics III, 3-50 May 2010, Cyprus. WIT Transactions on Ecology and the Environment.

Estimating the economic benefits of redeveloping the former Athens International Airport

D. Damigos & E. Laliotis
National Technical University of Athens, Greece

Abstract

This paper illustrates the results of a survey carried out in order to evaluate two alternatives suggested for the redevelopment of the former Athens International Airport, a project known as "Hellenikon Metropolitan Park". The first plan provides for a mixed-use park involving both green areas with cultural, sports and leisure facilities and commercial uses that will offer a financial return. The second plan proposes solely the development of a green park with light recreational facilities. In order to examine the situation from a social point of view, the effect of the proposed plans to the property prices in the surrounding area is examined by means of the Fuzzy Delphi Method. The results indicate that although both plans will positively affect the price of dwellings in a similar influence zone, the second plan will create a premium almost 60% higher, compared to the first one. Although a clear answer can only be obtained through further research, the findings could be used in order to justify the evaluation of the proposed land-use alternatives.

Keywords: Fuzzy Delphi Method, property values, brownfields redevelopment, urban green areas.

1 Introduction

Since the mid-1980s, there has been a growing recognition that brownfields (i.e. former industrial sites, airports, railway stations, etc.) should not be solely seen as a problem (e.g. due to pre-existing historic contamination) but also as an opportunity for enhancing local development in a sustainable manner. The policies implemented so far intend primarily to lessen the costs and risks

WIT Transactions on Ecology and the Environment, Vol 131, © 2010 WIT Press
www.witpress.com, ISSN 1743-3541 (on-line)
doi:10.2495/EEIA100041

associated with the regeneration of brownfields. As a result, the vast majority of remediation efforts focus mainly on converting brownfields into productive uses (i.e. commercial, residential or even industrial).

Nevertheless, the creation of green space by reclaiming brownfields could provide significant social, environmental and economic benefits, e.g. air pollution control, noise attenuation, improvement of microclimate, provision of recreational opportunities, etc. [1]. For this reason there exists a widening interest to support this type of development, especially within urban boundaries. For example, between 1988 and 1993, over 19% of brownfield sites in Britain were converted into green areas [2]. However, the costs of redeveloping the derelict land can be assessed in a straightforward way but the benefits of non-commercial uses, i.e. green areas, are hard to estimate in monetary terms. Thus, they are often overlooked in decision-making procedures. In order to confront this situation, the use of environmental economics has been proved to be beneficial, since several research efforts have established that green spaces provide monetary benefits in many different ways [3–6], and especially through the increase of surrounding property values [7–10].

This paper highlights the abovementioned issues through an illustrative case, namely the reuse of the former Athens International Airport. More specifically, two redevelopment alternatives of the so-called "Hellenikon Metropolitan Park" project are evaluated on the basis of results obtained by estimating the effect of the proposed plans to the property prices in the surrounding area. Towards this aim, the Fuzzy Delphi Method, a well-established group judgment technique, is applied in order to overcome some of the theoretical and practical complexities of the Hedonic Pricing analysis.

2 Methodological background

2.1 The hedonic approach

The hedonic pricing method is a well established approach for estimating the effect of environmental quality on housing values, especially in the urban setting. According to the theoretical concept of the method, dwelling prices differ with respect to housing characteristics (square footage, number of rooms, quality of accommodation, etc.), neighbourhood characteristics (level and quality of social infrastructure, housing density, traffic, etc.) and the quality of the environment (air quality, noise level, landscape features, etc.). Other characteristics being equal, it would be reasonable to expect that properties in areas with better environmental quality enjoy higher prices. The hedonic method has been used in a variety of applications in order to reveal the monetary value of environmental attributes such as: clean air, proximity to green areas and open spaces, view to lakes and forests, noise levels, etc., e.g. [7–9, 11, 12].

The application of the method prerequisites the availability of extensive cross-section data, time series data or a mixture of both concerning real prices and characteristics of dwellings in the area of interest, in order to isolate the contribution of the environmental factor to the market price by means of econometric techniques. In addition, data should be collected from a reasonably stable market period [13]. Further issues arise with respect to model specification and level of disaggregation, market distortions, etc. [13–15]. Finally, it should be mentioned that the method cannot be easily applied when potential changes on the environmental quality are investigated, thus it is mainly used in ex post analyses.

2.2 The Fuzzy Delphi Method

Due to the lack of adequate datasets, the instability of housing market in Greece during the previous years and the fact that the hedonic analysis cannot be applied in ex ante cases, an alternative approach was adopted, namely the Fuzzy Delphi Method (FDM).

The FDM is an analytical process based on the Delphi Method and the theory of fuzzy sets [16]. The theoretical assumptions and methodological procedures of the Delphi method were developed in the 1950s and 1960s at the RAND Corporation, at Santa Monica, California [17, 18]. The method is actually a structured process for the systematic collection and collation of judgments from a group of experts on a particular issue, by means of a series of questionnaires interspersed with controlled opinion feedback [19]. The experts are requested to give their opinion separately and independently about the variables in question. The results of the first round are analyzed statistically by finding their average and are then interspersed to the participants, who are asked if they wish to revise their earlier estimates. This process is followed again and again until the outcome converges to a reasonable solution from the point of view of the decision maker.

In order to deal with the effect of subjectivity of the experts, as well as the uncertainty imposed by the complexity of the problems studied, the theory of fuzzy sets, known also as fuzzy logic, is usually involved in the context of the FDM [20]. Fuzzy sets are an extension of the classical set theory. A fuzzy set is characterized by a membership-degree function, which maps the members of the Universe into the unit interval [0,1]. The value 0 means that the member is not included in the given set, 1 describes a fully included member. Hence, for the universe $\boldsymbol{U}$ a fuzzy set $\boldsymbol{A}$ is defined by as:

$$\boldsymbol{A} = \{x, \mu_A(x)) \mid x \in \boldsymbol{A}, \mu_A(x) \in [0,1]\}$$

where $\mu_A(x)$ is the membership-degree function μ: $x \rightarrow [0;1]$.

A fuzzy number is defined in the universe $\boldsymbol{R}$ as a convex and normalized fuzzy set. In this case, the triangular numbers were considered to be mostly suitable, since they could be constructed easily by asking the experts to specify three values, the minimum, the maximum, and the most plausible. More specifically, the triangulated fuzzy number $\boldsymbol{T}$ with membership function $\mu_A(x)$ is defined on $\boldsymbol{R}$, as follows:

$$T = \begin{cases} \frac{x-a}{b-a} & a \le x \le b \\ \frac{x-c}{b-c} & b \le x \le c \\ 0 & \text{otherwise} \end{cases}$$

where $[\boldsymbol{a},\boldsymbol{c}]$ is the supporting interval and the point $(\boldsymbol{b},\boldsymbol{1})$ is the peak.

The Fuzzy Delphi Method consists of the following steps [21]:

Step 1. The experts E_i, $i = 1,2,\ldots,n$, are asked to provide their estimates on the particular subject, determining the minimum $\alpha_1^{(i)}$, the most plausible $a_M^{(i)}$ and the maximum $a_2^{(i)}$. The data given by the experts Ei are presented in the form of triangular numbers:

$$A_i = (a_1^{(i)}, a_M^{(i)}, a_2^{(i)}),\ i = 1,2,\ldots,n$$

Step 2. The fuzzy average $A_{ave} = (m_1, m_M, m_2)$ of all A_i is estimated, according to the equation:

$$A_{ave} = (m_1, m_M, m_2) = \left(\frac{1}{n}\sum_{i=1}^{n} a_1^{(i)}, \frac{1}{n}\sum_{i=1}^{n} a_m^{(i)}, \frac{1}{n}\sum_{i=1}^{n} a_2^{(i)}\right)$$

Next, for each expert the deviation between A_{ave} and A_i, is computed, which is a triangular number defined by:

$$A_{ave} - A_i = (m_1 - a_1^{(i)}, m_M - a_M^{(i)}, m_2 - a_2^{(i)}) =$$
$$\left(\frac{1}{n}\sum_{i=1}^{n} a_1^{(i)} - a_1^{(i)}, \frac{1}{n}\sum_{i=1}^{n} a_m^{(i)} - a_m^{(i)}, \frac{1}{n}\sum_{i=1}^{n} a_2^{(i)} - a_2^{(i)}\right)$$

The deviation $A_{ave} - A_i$ is given back to the experts for revision.

Step 3. Each expert E_i gives a new triangular number:

$$B_i = (b_1^{(i)}, b_M^{(i)}, b_2^{(i)}),\ i = 1,2,\ldots,n$$

This process, starting with step 2, is repeated, until two successive means become reasonably close, according to the decision maker.

The Delphi approach has been criticized for dependency of forecasts on the particular judges selected, the sensitivity of results to ambiguity in the questionnaire and the difficulty in assessing the degree of expertise incorporated into the forecast [22, 23]. Nevertheless, several studies indicate high agreement between the Delphi estimates and the real numbers [24, 25].

3 Case study

3.1 The former Athens International Airport site

The former Athens International Airport, known as "Hellinikon International Airport" was built in 1938. The site, located 11.5 km south from the centre of Athens, covers an area of 530 hectares and borders residential areas, the Gulf of Saronikos and the Glyfada Golf Club (Fig. 1).

Figure 1: The former “Athens International Airport” (Source: SERERO architects).

In 1950 the airport had already equipped with two runways of 2,250 m in length. By the end of 50’s the Airport handled 500,000 passengers and 4,000 tonnes of cargo per year and the main runway extended to 3,500 m to accommodate traffic growth and jet aircrafts. Having predicted that the airport would handle up to 2,400,000 passengers annually by 1968, a new Terminal, known as the East Air Terminal, was built. The East Air Terminal, designed by the Finnish-American architect E. Saarinen, opened in 1969. In the early 70’s the Airport handled 3,300,000 passengers and 25,000 tonnes of cargo per year. In 1997 the number of passengers reached 12,000,000, while the cargo volume handled was 120,000 tonnes. The passenger traffic was increasing rapidly and it was evident that the Airport could not meet future demand since the surrounding residential area prohibited any further expansion. Thus, a new study was commissioned aimed at the relocation of the Airport. In March 2001, after an uninterrupted operation of 60 years, the Athens International Airport was finally moved to a new location, at Spata, where the new “Eleftherios Venizelos” Airport was built [25].

3.2 The redevelopment of the site

After its closure, a portion of the airport was redeveloped, hosting the venues for basketball, fencing, canoe/kayak slalom, field hockey, baseball and softball during the 2004 Summer Olympics. Nowadays, the “Hellenikon Complex”, covering an area of about 80 hectares, is managed by the “Hellenic Olympic Properties S.A.”, a state-owned enterprise. The complex currently comprises six sport grounds and two training facilities. An area of about 20 hectares in the East Air Terminal is administered by the “Hellenic Tourism Development Co.”, a state-owned enterprise, as well. Furthermore, an area of about 11 hectares is used as tram and bus depot and 5.5 hectares have been given away to local entities. Finally, the area includes 419 buildings, of which 126 belong to the Civil Aviation Authority. Although the northern and western portions of the site have changed, the Athens radar centre is located there and part of the old airport and its runways still exist at an area of about 300 hectares.

Since the termination of operations of "Hellinikon International Airport" in the end of March 2002 and especially after the 2004 Summer Olympics, there have been discussions concerning the reuse of airport's facilities. In June 2006, the Greek Minister for the Environment, Physical Planning and Public Works announced a final draft plan known as "Hellenikon Metropolitan Park", according to which 400 hectares will be converted to a park, while 100 hectares will be used to accommodate housing and office facilities.

It is referred that the proposed plan will create the largest urban park in Europe and will enhance the standards of living conditions of Athens' inhabitants. In total, green areas and amenity spaces will exceed 550 hectares, given that the plan also foresees: (a) the construction of an underground road tunnel for the adjacent Poseidonos Avenue that will allow the connection of the park with the beachfront and (b) the demolition of 378 out of the 419 buildings of the old airport. Regarding building development, the plan involves a business zone of 65 hectares and a residential zone of 35 hectares, respectively. Furthermore, a new museum of modern art is also planned. In total, it is estimated that built-up areas will cover 26 hectares, while the rest surface will be provided for public/common uses and infrastructure. It is envisage that the state will earn 500 million Euros from selling this area to land developers. This amount will fund the "Hellenikon Metropolitan Park" project, as well as the creation of smaller parks in devastated areas of Athens.

Nevertheless, representatives from municipalities surrounding the park and other entities accused the government of handing over a large chunk of privileged land to developers. In addition, they argued that the proposed plan, instead of the metropolitan park announced, will create a new city with a population of 15,000 - 20,000 people. Therefore, they formulated and counter proposed an alternative plan involving solely the development of a green park with light recreational facilities. This plan will establish additionally 30 – 100 hectares of green space (depending on the final design of the built area proposed by the first plan). On the other hand, the cost of the project will be totally covered by the state.

Considering that the city of Athens presents an unacceptable rate of proportional green space per capita, both plans will improve the quality of urban environment, though at a different level. From a social point of view, the critical question is whether or not 100 hectares of green space worth 500 million Euros that will be earned from developing an equivalent portion of the site. However, in order to answer this question the alternatives should be compared on an equal basis, i.e. the environmental and socio-economic benefits offered by the additional green space should be monetized.

3.3 Survey characteristics

In order to provide land-use planners and decision-makers with some useful pieces of information, the monetary benefits derived from the redevelopment alternatives were estimated by means of the Fuzzy Delphi Method. Towards this direction, a panel consisting of ten real estate experts, namely realtors, was employed. The participants' background in terms of professional skills and experience was taken into consideration, in order to maximize the effectiveness of the study. Panellists were provided with guidelines to increase the reliability of their answers and they were told that they were free to add their comments.

The experts were provided with a specially formed questionnaire, in order to give their estimates for the effect of each redevelopment alternative on the price of dwellings located in the vicinity. The questionnaire consisted of a comprehensive list of twenty eight questions. The first set included questions regarding the effect of the airport, while it was operating. More specifically, the experts were asked to state their estimations with respect to: (a) the effect of the airport on dwellings prices (positive or negative), (b) the influence zone of the airport (in km), and (c) the change in average dwelling price (in percentage) within the influence zone. The second set referred to the influence of the termination of operations of the "Hellinikon Airport" on the surrounding housing market, using a similar set of three questions, as described above. The third set of questions investigated the "announcement effect" of airport's development. The experts were provided with the same set of questions, in order to determine the effect of the announcement in terms of influence range and price alteration. The final set of questions focused on a hypothetical case. The panellists were asked to forecast the effects of each of the proposed alternatives on the housing market of the neighbouring area with respect to the zone of influence of each alternative on the dwelling price (in km) and the premium attracted by a typical dwelling located in the zone of influence (in percentage)

It should be noted that the answers given by the experts regarding the questions of the first three sets derived from market data, while those provided to the final set of the questionnaire were based on estimates. Given the scope of the paper, only the results referred to the last part of the questionnaire, i.e. the effect of the proposed alternatives, are presented in the next section.

3.4 Results

The survey was completed in two rounds, since the point of diminishing returns was considered to be satisfying. For conciseness reasons, only the results of the second round are presented.

According to the results, the Alternative A proposed by the state will influence, on average, the surrounding lodgings on a range up to 3 km and will modify their value between 11% - 33%, with the most probable rate being 23%. The implementation of Alternative B, which is counter proposed by local entities, is expected to affect the dwelling prices at the same range, i.e. up to 3 km. A dwelling within this zone, however, will attract, on average, a premium of

Table 1: Influence zone (in km).

E_i	A_{min}	A_{likely}	A_{max}	B_{min}	B_{likely}	B_{max}
E1	0	2	4	0	1	2
E2	0	2	3	0	2	3
E3	0	2	3	0	2	3
E4	0	1	2	0	1	2
E5	0	1	2	0	2	2
E6	0	2	3	0	2	2
E7	1	2	4	0	1	2
E8	0	1	2	1	2	3
E9	0	1	2	0	1	2
E10	0	1	2	0	1	2
Average	***0***	***2***	***3***	***0***	***2***	***3***
Minimum	***0***	***1***	***1***	***0***	***1***	***2***
Maximum	***1***	***2***	***4***	***1***	***2***	***3***

Table 2: Price alteration (in percentage).

E_i	A_{min}	A_{likely}	A_{max}	B_{min}	B_{likely}	B_{max}
E1	10	20	30	15	35	50
E2	10	20	30	25	35	60
E3	10	30	35	25	30	40
E4	20	35	50	20	30	40
E5	5	15	20	20	50	70
E6	5	15	30	15	30	50
E7	10	20	30	20	30	50
E8	15	25	40	20	35	40
E9	15	25	35	20	40	60
E10	10	20	30	25	40	60
Average	***11%***	***23%***	***33%***	***21%***	***36%***	***52%***
Minimum	***5%***	***15%***	***20%***	***15%***	***30%***	***40%***
Maximum	***20%***	***35%***	***50%***	***25%***	***50%***	***70%***

21% up to 52%, and most probably up to 36%. The results indicate that although both plans will positively affect the price of dwellings in a similar range, the second plan will create a premium almost 60% higher, compared to the first one.

Given that in year 2006 the average unit price of a dwelling in the area of interest was 3,000 Euro/sq.m, the added value to the properties is estimated to 690 Euros/sq.m and 1,080 Euros/sq.m for the first and the second alternative, respectively. These figures amount to 69,000 Euros and 108,000 Euros, correspondingly, for a typical apartment in the surroundings (saleable area: 100 sq.m.).

4 Discussion

Taking into account the findings of the study, it would be logically assumed that the monetary benefits attributed to the additional green space are not sufficient to justify the trade-off of selling the area to land developers. Given that the difference in the additional premium attracted by a dwelling between the two alternatives is 39,000 Euros, the second plan should influence approximately 13,000 apartments. This figure certainly exceeds the total number of residences located within the influence zone. Furthermore, it should be also considered that this added value could be detached only by means of a special property tax in a depth of time.

On the other hand, however, the economic benefits of urban forests are not restricted to the increase in property value. For example, the energy savings of buildings due to the impact of urban parks to the microclimate are also significant, considering that savings on cooling costs for a typical household have been estimated at 1.9% to 2.5% per residential tree [3]. McPherson [26] estimated that an urban park of 210 hectares in Chicago provided air pollution reductions equivalent to traditional emission controls costing 136 USD per day, while another study indicated that the annual aggregate value of stormwater management of the existing tree cover of the USA cities, in 1997, was 400 billion USD [27]. In addition to those findings, several studies indicate significant use and non-use values associated with urban forests. In Finland, for instance, researchers found that households were willing to pay an amount of 14.4-27.2 Euros per year in order to prevent the conversion of urban forests to another land use [4]. In Spain, people were willing to pay an annual tax of 71 Euros in order to construct an urban park of 28 hectares in the centre of Valencia, on the site of an old train station [28]. In the city of Athens, households were willing to pay an annual fee of 41.5 Euros in order to establish a forestry organization for maintaining and enhancing city's green spaces [29].

Urban parks and green spaces are subject to development pressure because their benefits are hard to estimate in economic terms. In the case of "Hellenikon Metropolitan Park", the results definitely prove that even a 'pure' green area would create significant economic value for the society. Thus, although a final answer to the critical question can only be gained through further research, these findings should be taken into account in order to come up with a more sound and socially fair solution with respect to the proposed land-use alternatives.

References

[1] Commission of the European Communities (CEC), *Green Paper on the Urban Environment*, COM(90) 218 Final, Brussels, 27 June, 1990.

[2] De Souza, C., Turning brownfields into green space in the City of Toronto, *Landscape and Urban Planning*, **62(4),** pp. 181-198, 2003.

[3] Simpson, E.G. and McPherson, J.R., Simulation of Tree Shade Impacts on Residential Energy Use for Space Conditioning in Sacramento, *Atmospheric Environment*, **32(1),**pp. 69-74, 1998.

[4] Tyrväinen, L. and Väänänen, H., The economic value of urban forest amenities: an application of the contingent valuation method, *Landscape and Urban Planning*. **43(1-3)**, pp. 105–118, 1998.
[5] Damigos, D. and Kaliampakos, D., Assessing the benefits of reclaiming urban quarries: a CVM analysis, Landscape and Urban Planning, **64(4)**, pp. 249-258, 2003.
[6] Nowak, D.J. and Dwyer, J.F., Understanding the benefits and costs of urban forest ecosystems. In: Kuser, J., ed. *Urban and community forestry in the Northeast, 2nd edition*, Springer, the Netherlands, pp. 25 – 46, 2007.
[7] Anderson, L.M., and Cordell, H.K., Influence of trees on residential property values in Athens, Georgia (U.S.A.): A survey based on actual sales prices, *Landscape and Urban Planning*, **15(1-2)**, pp. 153-164, 1988.
[8] Tyrväinen, L. and Miettinen, A., Property prices and urban forest amenities, *Journal of Environmental Economics and management*, **39**, pp. 205-233, 2000.
[9] Luttik, J., The value of trees, water and open space as reflected by house prices in the Netherlands, *Landscape and Urban Planning*, **48(3-4)**, pp. 161-167, 2000.
[10] Damigos, D. and Kaliampakos, D., Forecasting the effects of environmental changes on residential land prices: An application of Fuzzy Delphi Method, *Thirteenth Annual Conference of EAERE*, June 25-28th, Budapest, Hungary, 2004.
[11] Rosen, S., Hedonic prices and implicit markets: product differentiation in pure competition, *Journal of Political Economy*, **82**, pp. 34-55, 1974.
[12] Nelson, J.P., Highway noise and property values: a survey of recent evidence. *Journal of Transport Economics and Policy*, **XIC**, pp. 37-52, 1982.
[13] Kula, E. *Economics of Natural Resources, the Environment and Policies*, Chapman and Hall, London, U.K., 1994.
[14] Butler, R.V., The specification of housing indexes for urban housing, *Land economics*, **58**, pp. 96 – 108, 1982.
[15] Palmquist, R.B., Hedonic methods. In: *Measuring the demand for environmental quality*, Braden, J.B. and Kolstad, C.D. eds. North-Holland, Amsterdam, pp. 77-120, 1991.
[16] Zadeh, L., *Fuzzy Sets and Applications: Selected Papers*. John Wiley & Sons, New York, 1987.
[17] Dalkey, N. and Helmer, O., *An Experimental Application of the Delphi Method to the Use of Experts*, The RAND Corporation, Santa Monica, California, USA, 1962.
[18] Dalkey, N., The Delphi Method: *An Experimental Study of Group Opinion*, RM-5888-PR, The RAND Corporation, Santa Monica, California, USA, 1969.
[19] Adler, M. and Ziglio, E., *Gazing into the Oracle: The Delphi Method and its Application to Social Policy and Public Health*, Philadelphia: Taylor and Francis, 264 p., 1996.

[20] Kaufman, A. and Gupta, M.M., *Fuzzy Mathematical Models in Engineering and Management Sciences*, North-Holland, Amsterdam, 1988.
[21] Bojadziev, G. and Bojadziev, M., *Fuzzy Logic for Business, Finance and Management*, Word Scientific Publishing, Singapore, pp. 22-23, 1997.
[22] Sackman, H., *Delphi Assessment: Expert Opinion, Forecasting, and Group Process*, The RAND Corporation, Report #R-1283-PR, Santa Monica, California, USA, 1974.
[23] Madridakis, S. and Wheelwright, S.C., Interactive Forecasting, *Univariate and Multivariate Methods*, Second Edition. Holden-Day. San Francisco, 1978.
[24] Milkovich, G., Anthony A. and Thomas M., *The Use of Delphi Procedures in Manpower Forecasting*, Management Science, **19(4)**, pp. 381-388, 1972.
[25] Air Traffic Safety Electronic Engineers Association of Hellenic Civil Aviation Authority, Athens International Airport History, http://www.hcaa-eleng.gr/athhist.htm#English
[26] McPherson, E.G., Environmental benefits and costs of the urban forest, in Rodell. P.D., (ed.) *Proceedings of the Fifth National Urban Forest Conference, Los Angeles*, Washington, DC, American Forestry Association, pp. 52-54, 1991.
[27] American Forests, *The State of Our Urban Forests: Assessing Tree Cover and Developing Goals*, White paper Washington, D.C., 1997.
[28] del Saz Salazar S. and Menéndez G, *Estimating the non-market benefits of an urban park: Does proximity matter*? Land Use Policy, **24(1)**, pp. 296-305, 2007.
[29] Kalavrytinos, N. and Damigos, D., *The Economic Value of Urban Green Spaces in the Attica Basin*, Technica Chronica Scientific Journal, Technical Chamber of Greece, **II, 1-2**, pp. 7-21, 2006.

Assessment of the impact of local energy policies in reducing greenhouse gas emissions

A. Arteconi, C. M. Bartolini, C. Brandoni & F. Polonara
Department of Energy, Polytechnic Marche, Italy

Abstract

The present work investigates the potential energy savings coming from a careful and detailed local energy policy. The paper analyzes and assesses from technical, economic and environmental viewpoints different initiatives in the energy sector aimed at increasing energy efficiency in end-uses and reducing overall carbon emissions. The results are based on energy planning for five Italian urban areas sized at about fifty thousand inhabitants. The analysis has been developed with the aim of reaching generally applicable criteria suitable for evaluating the local energy policy contribution to the greenhouse gas (GHG) emissions reduction. Several initiatives for the private and public sectors have been considered, such as: (i) the introduction of combined heat and power generation based on useful thermal demand, suitable for the industry and service sectors (swimming pools, large distribution organizations); (ii) generation of electricity from renewables (solar energy, biomass); (iii) thermal insulation of private and public buildings, such as schools; (iv) introduction of micro-combined heat and power generation in the residential sector. For each solution the primary energy reduction and the consequent reduction in GHG emissions have been evaluated and a feasibility analysis has been developed in order to assess the profitability of the investment. Great attention has been paid to the public sector, which has an important role in providing leadership and driving changes in other sectors; furthermore, a rational use of energy combined with the exploitation of country-based incentives is able to reduce the public administrative expenditure. The results show that local energy policy can give an important contribution to gas emission reduction targets, and underline the fundamental role of public sector initiatives.

Keywords: municipal planning, energy policy, renewable, energy savings, micro-CHP.

WIT Transactions on Ecology and the Environment, Vol 131, © 2010 WIT Press
www.witpress.com, ISSN 1743-3541 (on-line)
doi:10.2495/EEIA100051

1 Introduction

Copenhagen United Nations Climate Change Conference ended in December 2009 with a political agreement that recognized climate change as one of the greatest challenges of our time. It was decided to cap temperature rise under two degrees, to reduce emissions and to raise finance for actions aimed at combating climate change [1]. The Copenhagen Accord can be considered as the first step that involves most countries and gives basis to an international operative program, but it was not legally binding for the parties. In order to activate GHG emission reduction initiatives, most countries understand the importance of strategic energy planning not only on a national scale, but also on regional and local levels [2]. Even in the US [3], which first rejected mandatory targets for curbing emissions under the Kyoto Protocol, lots of actions have been undertaken in the energy sector at the local level, in order to fill the federal leadership vacuum. In recent years many urban areas have developed municipal energy planning, adopting an environmental sustainability policy to improve living conditions and environments [4]. Several discussion networks were born in Europe and abroad with the aim of sharing energy savings 'bottom-up' initiatives and assessing the potential of local energy policy instruments. Furthermore, several initiatives were undertaken, as Local Agenda 21, to assist local authorities in Climate Protection Campaigns. The need to catch the economic opportunities of investment in energy efficiency and renewables, in order to reduce the administrative expenditure, is another important factor emphasizing the importance of local energy policy. As a matter of fact, by encouraging local energy policy it is possible to create jobs and generate revenue for the local economy. In Italy the importance of municipal energy planning was first outlined with Italian national law no.10/91, which defined the basis for national Energy Planning development and was focused on developing local energy programs for the rational exploitation of renewable resources for urban areas with more than 50 thousands inhabitants. In recent years funds have been allocated to develop municipal energy programs in order to increase energy efficiency in end-uses, and adopt a rational exploitation of renewables. In the case under study the Regional Government provided a loan to municipalities characterized by having about fifty thousand inhabitants, in order to develop environmental and energy programs with the aim of integrating energy into the local policy.

2 Methodology

The results are based on five municipal energy and environmental plans of urban areas located in central Italy with about fifty thousand inhabitants. Table 1 shows the number of citizens registered in 2008. The analysis has been developed in order to reach generally applicable criteria suitable to evaluate the local energy policy contribution to the greenhouse gas (GHG) emissions reduction. Each energy municipal plan is made up of three phases: (i) an *investigative phase* that analyzes the local energy supply and demand in order to point out criticalities in energy production and end-users, (ii) an *analysis phase* that studies the actions to improve a rational use of energy and renewables, (iii) an *executive phase* where the potential GHG emission reduction is assessed. In all of the analyses of energy supply and demand, the public sector has been considered independently.

The public sector actually has an important role in providing leadership and driving changes in other sectors. In the *analysis phase* each initiative has been analyzed from both energy and economic viewpoints. When possible, a standard action and the simple pay back method has been defined, eqn. (1), and has been used to assess the economic profitability of the investment, based on the calculation of the initial investment, Ex, and on annual cash flow, F, generated by the energy-saving solution.

$$PBP = \frac{Ex}{F} \tag{1}$$

In the *executive phase*, firstly, the potential reduction in primary energy derived from the energy saving actions and its real feasibility on the basis of factors derived from literature and the author's experience has been assessed. Secondly, the GHG emission reduction has been calculated using emission factors. Furthermore, in order to generalize results, some generalized factors, suitable for estimating the contribution of the local 'bottom-up' initiative, have been found.

3 Investigative phase

Figures 2 and 3 show the percentage sector shares in total electric and thermal consumption for the five urban areas analyzed.

Table 1: Number of citizens in the urban areas under study.

urban areas	1	2	3	4	5
inhabitants	44,207	93,488	41,056	51,503	63,734

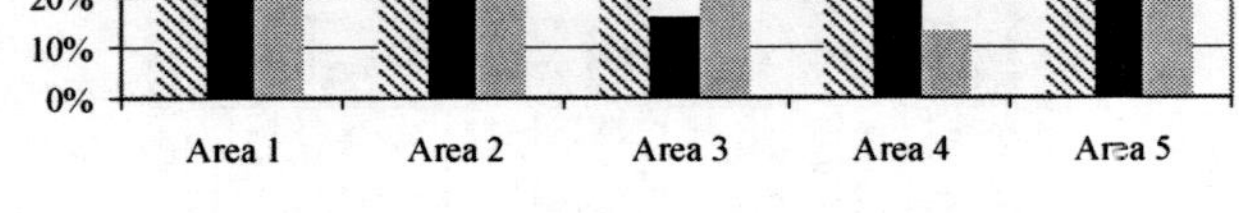

Figure 1: Electric consumption.

Except for the area No. 4, which is characterized by a high incidence of the industry sector, due the presence of energy intensive industries, such as paper industry, and to a smaller extent the area No. 5, the service sector is the most critical one in the total electric consumption.

The energy consumption of the service sector comprises the energy used in private and public buildings. Due to the tertiarization of the economy the service sector incidence in energy consumption is becoming more and more important. As far as the thermal demand is concerned, the most critical activity sector is the residential one.

The number of dwellings by year of construction has been analyzed in order to understand the reason why the residential sector is so energy intensive and define the best energy-efficiency measures (figure 4). More than the 80% of dwellings in all the areas analyzed have been built up to 1991, before the first law aiming at improving thermal performances was introduced in Italy. Another important result coming from the investigation phase is the low incidence of renewable energy production for all the areas analyzed, on average equal to 2%. The result suggests the important contribute in terms of GHG emission reduction, which could come from the introduction of renewables.

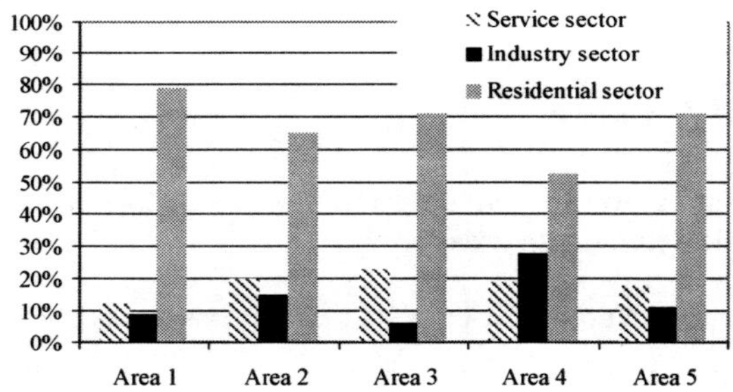

Figure 2: Thermal consumption.

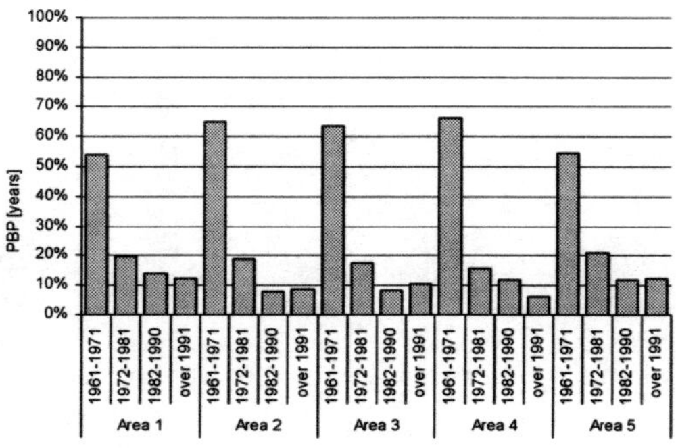

Figure 3: Number of dwellings by year of construction.

Eventually, the public sector has been analyzed in detail with the aim of planning important energy saving initiatives to reduce the public sector energy bill and to drive energy saving actions. Graph 5 shows the electric and thermal energy consumption of schools, offices and sport facilities in the five areas. Data collected suggest several actions which can be adopted in order to maximize the energy bill reduction, such as thermal insulation in schools which most affect the municipal thermal consumption, the introduction of electric efficiency equipment in offices, which reveals to have a big impact in the electric consumption and the introduction of CHP plants in sport facilities.

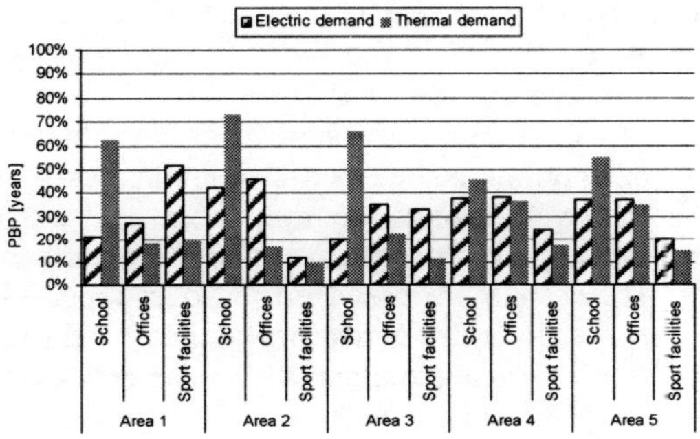

Figure 4: Energy consumption in the public sector.

4 Analysis phase

This paragraph deals with significant and repeatable actions studied for the main activity sectors with a focus on the energy measures for municipal sector.

4.1 Residential sector

Several initiatives were studied for the residential sector: (i) thermal insulation actions for dwellings built before 1991, (ii) adoption of Italian law 192/2005 in the existing municipal 'building code' aiming at reducing the primary energy building consumption, (iii) information campaign for high-energy efficiency household equipment, (iv) replacement of electric boilers with solar collector systems, (v) introduction of micro-CHP devices. As mentioned in the analysis phase the residential sector is the main responsible for thermal consumption, mainly due to a lack of accurate thermal insulation building design. Although the procedure to reduce the thermal losses implies an accurate survey of thermo-physical performances of the building envelope, a standard action has been analyzed in order to assess its potential contribution to clean air policy. The standard thermal insulation action is referred to an apartment of 100 square meter area, built before 1991; the energy savings have been calculated applying the best available techniques and materials. It has been assessed a primary energy reduction of about 35%, the resulting payback period, considering the revenue deriving from the reduction in energy bill, is of about twelve years (Table 2). The second energy saving measure analyzed addresses new buildings. The contribution coming from the adoption of Italian law Decree 192/2005 in the existing municipal 'building code' has been assessed. The Italian Legislative Decree n. 192 (and its corrections and integrations by Legislative Decree 29th December 2006, n. 311) brought into force the Directive 2002/91/CE on energy efficiency in buildings. The Decree aims at improving buildings energy performances strengthening the thermal insulation standards. Furthermore the Italian law established additional measures such as the satisfaction of more than 50% of hot water demand by solar generation and compulsory installation of 1 kWel photovoltaic panels for each new dwelling. The primary energy consumption for a unit property in the five areas analyzed is, on average, 1.1 toe/year. The introduction of law 192/2005 will decrease the total energy consumption down to 0.75 toe/year, on average. Another important item in residential energy consumption is the electric equipment used in dwellings. It has been assessed the energy savings coming from the introduction of high efficiency household equipment driven by an information campaign developed by municipalities. The results are based on the data reported in literature and shown in table 3. Another interesting measure studied for the residential sector is the replacement of electric water boilers with solar collector systems. In fact, although in the areas analyzed an extensive natural gas pipeline network makes the building sector dependent on this source for space heating, tap water heating and cooking, it has been found out that a small percentage of dwellings use electric water boilers. Table 2 sums up the costs connected to the action and the relative payback period. It has been supposed to apply flat-plane solar collectors characterized by an energy-efficiency of 32% when solar radiation is of 4.6 kWh/m^2/day. Eventually it has been evaluated the introduction of micro-Combined Heat and Power, micro-CHP, generation in residential sector, which can be considered one of the most innovative and interesting solution due to its high potential. Micro-cogeneration technologies are expected to play an important role in greenhouse gas emission reduction, as they ensure reliable energy production and may become competitive quite soon [5].

The European Parliament has recognized cogeneration and micro-cogeneration as efficient technologies capable of providing energy savings and of helping achieve climate policy objectives [6]. The market for most micro-cogeneration technologies is still immature, largely due to technical and

Table 2: Energy and economic assessment of energy-savings measures analyzed in the residential sector.

	Investment cost (€)	Energy savings	PBP (years)
thermal insulation	25-35,000	35%	12
electric boiler replacement with solar collector systems	800 €/m^2	0,014 toe/m^2/year	4
micro-CHP	12,000 €	15%	7

non-technical barriers, e.g. cost of the investment, red tape, suboptimal product quality (required to gain market trust), and lack of information (essential for the spread of the technology). Moreover little research has been done on the long-term cost and benefits of the technology; for instance the investment required to adapt transmission and distribution networks to its widespread use has not been clearly addressed [7]. Several combined heat and power systems are suitable for residential sector applications [8], i.e. reciprocating internal combustion engines, micro-turbines, fuel-cells and Stirling engines, from an economic point of view internal combustion Engine perform better. A detached house was considered as a 'standard case', it was assumed to apply a 5 kWel ICE; results shown in table 2 come from a previous study of the present authors [9].

Table 3: Energy savings derived from the introduction of high efficiency household equipment.

	Refrigerator	Washing machine	Dish washing machine	Freezer	High efficiency lamps
energy savings (toe/year)	0.019	0.009	0.008	0.022	0.0146
percentage number per dwelling	110%	37%	100%	25%	435%
high efficiency equipment installed so far	41.7%	38.9%	32.2%	32.2%	25%

4.2 Service sector

The service sector comprises buildings of private and public ownership. The private sector is characterized by different end-users in terms of energy demand, such as offices, hotels, supermarket, suitable for different energy-saving initiatives. It has been chosen to analyze two significant and repeatable measures: (i) the introduction of CHP plant in private swimming pools, which can be repeated for municipal swimming pools, (ii) the introduction of distributed generation system and polygeneration in supermarkets. Swimming pools are really suitable for the introduction of combined CHP systems due to an important need of thermal energy during the whole year. The energy demand for a swimming pool in one of the area analyzed has been taken as 'standard case' in order to assess the profitability of the investment. The hypothesis was to introduce a 100 kWel microturbine connected in parallel to the grid. Microturbines technology represents, nowadays, a potential alternative to ICE [10], particularly in the context of mini-cogeneration, by comparison with ICE, though they have a lower electrical efficiency and a higher initial cost, they are more compact and lightweight, they require less maintenance and they have a longer working life, thanks to a more straightforward architecture. Results are shown in table 4. Supermarkets are characterized by a strong demand for energy for refrigeration for food preservation and for ambient air-conditioning during the summer. This makes supermarkets particularly suitable for trigeneration applications with the prime mover coupled with absorption systems. Based on results of a study developed by the authors [11], it has been assumed to introduce trigeneration systems for the combined production of electricity and ambient heating and air-conditioning energy combining it with photovoltaic systems.

The analysis has been developed on the basis of the consumption of a typical supermarket characterized by a sales area of 10,000 m^2. Results are shown in table 4. In the public sector the initiatives analyzed are: (i) street lighting energy saving measures, (ii) thermal insulation in schools and (iii) installation of photovoltaic panels, PV, in public areas. Furthermore it has been assumed to introduce CHP in municipal swimming pool as previously introduced. The street lighting energy saving measures concern (i) the replacement of high pressure sodium (HPS) lighting with LED technology for outdoor application, and (ii) the replacement of incandescent lighting with LED technology in traffic lights systems. LED technology is characterized by several benefits such as longer lifetime, less maintains, high color brightness and efficacy. Tables 5 and 6 compare, respectively the sodium high pressure technology and the incandescent lamp with LED technology.

As in the residential sector, a standard thermal insulation action has been planned for schools; the results are reported in table 7. It has assessed a 30% reduction of primary energy consumption. The last initiative analyzed is the introduction of PV panels in public area. Solar energy is a strategic source for the areas analyzed. The problem connected with the introduction of PV energy in the

Table 4: Energy and economic assessment of energy-savings measures analyzed in the service sector.

	Investment cost (€)	Energy savings	PBP (years)
CHP in swimming pools	160,000	60 toe/year	7
DG & polygeneration in supermarkets	3,500,000	40%	9

Table 5: Comparison between HPS and LED technology.

		HPS	LED
lamp power	(W)	150	40
lifetime	(hours)	10,000	100,000
operational hours	(hours/year)	3,500	
energy saving for each replacement	(toe/year)	0.07	

Table 6: Comparison between incandescent lamp and LED technology.

		HPS	LED
lamp power	(W)	70	13
lifetime	(hours)	5,000	100,000
operational hours	(hours/year)	2,900	
energy saving for each replacement	(toe/year)	0.03	

public sector is the high investment required to the municipality. A possible solution is to let public areas to a private company which recognizes a fee to the municipal administration for a negotiated number of years. Another possible solution is an agreement with an Energy Service Company, ESCo; in this case the municipal institution can keep its energy costs constant and the ESCo retains ownership over the equipment until the end of the contract. During the contract duration the municipality is paying the same energy costs, but in the form of a service fee to the ESCo, in this service fee also included equipment modernization which is provided by the ESCo [12].

This is an opportunity to modernize the municipal equipment. Table 7 shows the parameters considered in the evaluation of the energy savings related to the PV initiative.

4.3 Industry sector

The introduction of distributed generation, DG, systems has been studied in order to make the industry sector as self-sufficient as possible. Using DG systems [13] offers a number of advantages, such as a reduction in the energy costs for the users and for the domestic and international economies as a whole, fewer losses in transmission, fewer carbon dioxide emissions, a better quality electrical energy generation and a less vulnerable electrical system. CHP plants have been assessed in the condition of contemporary demand of electric and thermal energy. It has been analyzed the possibility to feed the CHP plants with natural gas or biomass, particularly interesting for small plant (1-2 MW_{el}). Biomass is a renewable energy and it proves an interesting GHG emission reduction in case of 'short-chain' biomass plants. It means that the maximum distance between biomass production and usage must be within 50 kilometers.

Table 8 shows CHP investment cost per kW of electric energy installed; the energy savings and the relative PBP has been reported with reference to a 'typical' case. In case of biomass fuel it can be seen a reduction in the payback period derived from the national incentive of green certificate mechanism, which is a market mechanism introduced in Italy for incentivizing power generation from renewable sources.

Table 7: Energy and economic assessment of energy-saving measures in the public sector.

	Investment cost (€)	Energy savings	PBP (years)
thermal insulation in schools	50,000	33%	12
PV panel in public area	/	0.187 toe/kWh	10

Table 8: Energy-economic assessment of CHP in the industry sector.

	Investment cost (€/kW)	Energy savings (toe/year)	PBP (years)
natural gas feeding	1,200 €	9,000	7
biomass feeding	2,000 €	13,000	5

5 Executive phase

The potential primary energy reduction, derived from the energy saving actions analyzed, has been assessed and it has been evaluated its real feasibility on the basis of factors deduced from literature and author's experience (Table 9). A 5 year time period has been considered to take up all the initiatives developed. The results show the annual savings and emission reduction achievable since 2015. The GHG emission reduction has been calculated using an emission factor of 3 ton CO_2/toe, on the basis of the actual savings previously assessed. As far as the service sector is concerned it has been assumed that, at least, a CHP plant will be installed in a supermarket and in a private or municipal swimming pool by 2015. The results emphasize the important contribution in GHG emission reduction coming from municipal energy policy, a big contribution deriving from industry residential sector. As far as the service sector is concerned, more than half contribution come from the public sector, which, furthermore, can drive changes in the other sectors, especially in dwellings, through good practices.

Eventually, some generalized factors (table 10) have been found out in order to assess the municipal policy coming from the residential sector, an important aspect comes from the fact that each datum refers to one average dwelling.

Table 9: Emission reduction derived from the implementation of the energy savings initiatives.

	Potential savings (toe/year)	Factor (%)	Actual savings (toe/year)	GHG reduction (tonnCO_2/year)
Residential sector				
Thermal insulation	73,689	30%	22,107	66,320
Electric boiler replacement	4,456	40%	1,782	5,347
High efficiency lamp	6,044	50%	3,022	9,066
High efficiency household equipment	3,279	65%	2,131	6,393
Building code		/	540	1,620
Micro-CHP	23,487	5%	1,174	3,523
Service sector				
CHP in swimming pool	/	/	755	2,265
DG in supermarket	/	/	3,000	9,000
Public lighting measures	/	/	1,799	5,397
Thermal insulation in schools	/	/	1,787	5,361
PV panel in public areas	/	/	1,665	4,995
Industry sector				
DG system			35,000	105,000
Total			74,763	224,288

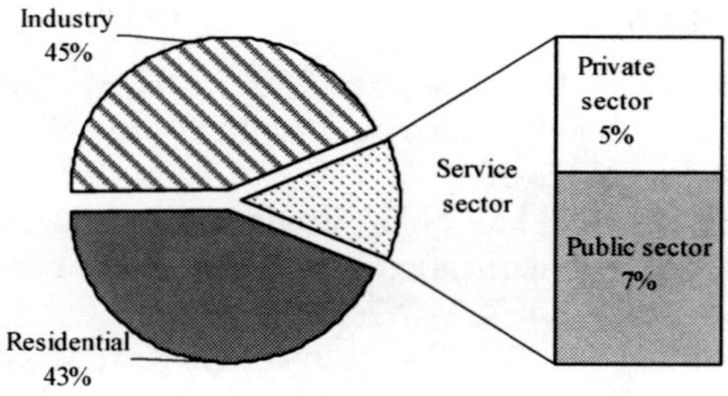

Figure 5: GHG emission reduction share in the five urban areas analyzed.

Table 10: Generalized factors suitable to assess the GHG emission reduction contribution derived from municipal energy planning.

	toe/apartment	tonnCO_2/apartment
thermal insulation	0.16	0.48
electric boiler replacement with solar collector systems	0.013	0.039
high efficiency lamp	0.023	0.069
high efficiency household equipment	0.016	0.048
efficiency building criteria in 'building code'	0.004	0.012
micro-CHP	0.009	0.027

6 Conclusion

The present paper analyzes and assesses from technical, economic and environmental viewpoints different initiatives in energy sector aiming at increasing energy efficiency in end use and reducing overall carbon emissions. Results are based on local energy planning studies for five Italian urban areas with about fifty thousand inhabitants. The analysis has been developed in order to reach generally applicable criteria suitable to evaluate the local energy policy contribute to the greenhouse gas (GHG) emissions reduction. The results point out the important contribution in GHG emission reduction coming from municipal institution, and the important role played by the municipalities in driving energy saving initiatives.

Acknowledgements

The authors are gratefully indebted to the five municipalities, in the Marche Region of Italy, which were the object of the analysis: Pesaro, Fano, Senigallia, Ascoli Piceno, San Benedetto.

References

[1] Copenhagen Accord, 18 December 2009, United Nations Framework Convention on Climate Change. Online. http://unfccc.int/resource/docs/2009/cop15/eng/l07.pdf

[2] R.B. Hiremath, S. Shikha, N.H. Ravindranath, Decentralized energy planning; modeling and application—a review, *Renewable and Sustainable Energy Reviews,* **11(5)**, pp. 729–752, 2005

[3] J. Byrnea, K. Hughesa, W. Rickersona, L. Kurdgelashvilia, American policy conflict in the greenhouse: Divergent trends in federal, regional, state, and local green energy and climate change policy, *Energy Policy,* **35(9)**, pp. 4555–4573, 2007.

[4] J. Stenlund Nilsson, A. Martensson, Municipal energy-planning and development of local energy-systems, *Applied Energy,* **76(1-3)**, pp.179–187, 2003.

[5] A.D. Hawkes, M.A. LeachCost-effective operating strategy for residential micro-combined heat and power, *Energy*, **32(5)**, pp. 711–732, 2007.

[6] H.I. Onovwiona, V.I. Ugursal, Residential cogeneration systems: review of the current technology, *Renewable and Sustainable Energy Reviews*, **10(5)**, pp. 389-431, 2006.

[7] M. De Paepe, P. D'Herdt, D. Mertens, Micro-CHP systems for residential applications, *Energy Conversion and Management*, **47(18-19)**, pp. 3435–3446, 2006.

[8] V. Kuhn, J. Klemesˇ, I. Bulatov, MicroCHP: Overview of selected technologies, products and field test results, *Applied Thermal Engineering*, **28(16)**, pp. 2039–2048, 2008.

[9] A. Arteconi, C. M. Bartolini, C. Brandoni, Energy and economic analysis of small-scale distributed generation in the residential sector, *Proc. of the 14thInt. Conf. On Stirling Engine,* 2009.

[10] J. Kaikko, J. Backman, Technical and economic performance analysis for a microturbine in combined heat and power generation, *Energy,* **32(4)**, pp. 378–387, 2007.

[11] A. Arteconi, C. Brandoni, F. Polonara, Distributed generation and trigeneration: Energy saving opportunities in Italian supermarket sector, *Applied Thermal Engineering*, **29(8-9)**, pp. 1735–1743, 2009.

[12] S. Rezessya, K. Dimitrov, D. Urge-Vorsatz, S. Baruch, Municipalities and energy efficiency in countries in transition: Review of factors that determine municipal involvement in the markets for energy services and energy efficient equipment, or how to augment the role of municipalities as market players, *Energy Policy,* **34(2)**, pp. 223–237, 2006.

[13] G. Chicco, P. Mancarella, Distributed multi-generation: A comprehensive view, *Renewable and Sustainable Energy Reviews,* **13(3)**, pp. 535–551, 2009.

The contradiction between modernising irrigation and water buyback

L. Crase & S. O'Keefe
Regional School of Business, La Trobe University, Australia

Abstract

The formulation of Australian water policy continues to be one of the most politically vexing areas of public policy. The pressing need for a national policy response stems from a variety of sources, including increasing scarcity, evidence of climate change, jurisdictional ranging amongst states and a history of generous allocations to agriculture that are now proving unsustainable. Collectively, these forces have promoted a policy stance that embodies serious contradictions. On the one hand substantial public investments are being made to upgrade irrigation infrastructure on the grounds that this will 'save' water that can then be allocated to satisfy environmental demands. On the other hand, governments have been actively repurchasing water rights, in the hope of striking a better balance in severely stressed systems, such as the Murray-Darling Basin. This paper traces the contradictions in this policy approach and argues that greater emphasis on the market purchase of water rights would give rise to a more efficacious outcome.
Keywords: water policy, water rights, market based instruments, irrigation.

1 Introduction

Until the late twentieth century, Australian governments, like many other administrations, viewed the development of water resources as a priority and saw the resource itself as a source of national prosperity [1]. Accordingly, generous allocation of water, particularly for agricultural pursuits, was seen as a precondition to food security, increasing national product and achieving social cohesion via the development of a noble yeomanry [2]. However, achieving these objectives against the backdrop of Australia's hydrology and climate proved particularly challenging. In simple terms, the variability of Australian

WIT Transactions on Ecology and the Environment, Vol 131, © 2010 WIT Press
www.witpress.com, ISSN 1743-3541 (on-line)
doi:10.2495/EEIA100061

climate and the factor endowments of the nation meant that irrigated agriculture was always destined to struggle on economic grounds alone [3].

In the 1990s these economic forces combined with greater concern for fiscal accountability and emerging environmental interests in the 1990s to usher in two major reforms in water policy [4]. The first was the Cap on water extractions collectively agreed by signatories to the Murray-Darling Basin Agreement. The ambition of the Cap was to limit the growth of surface water extractions to existing levels of development. The second was the introduction of the Council of Australian Governments' (CoAG) water reform agenda. An important component of this agenda was that water rights were to be separated from land and made tradeable [5, 6].

Perhaps ironically, an inherent conflict arose from these two policies. The source of this conflict was the large number of claims on water resources that were in existence at the time. Moreover, the decision by most state jurisdictions was to honour three main rights: statutory claims with a history of use; statutory claims with no history of use, and; non-statutory claims with evidence of a history of use. Regrettably, the quantum of water attending these claims was fundamentally in excess of the Cap. Moreover, the Cap itself is often seen as grossly inadequate for dealing with the longer term ecological sustainability of the river system [5].

Addressing the over-allocation of water resources in the Murray-Darling Basin is thus amongst the most pressing and difficult policy dilemmas facing governments. In this brief article we position the current policy stance and reflect on the inherent weaknesses and contradiction embodied in the status quo. More specifically, the contradiction between public investments to 'modernise irrigated agriculture' and the buying back of water rights is analysed. The paper concludes with a brief assessment of the practical means of overcoming (or at least limiting the impacts of) this contradiction.

2 Dealing with over-allocation

Over-allocation refers to situations where, with full development of entitlements in a particular system, the total volume of water able to be extracted by entitlement holders at a given time exceeds the environmentally sustainable level of extraction for that system [6]. In his highly acclaimed article on the environmental economics of the Murray-Darling Basin, John Quiggin [4] notes that there are four rudimentary policy choices for dealing with an over-allocation problem of this form. These comprise:

Government purchase of water rights (buy-back)

Allowing rights to degrade over time (say, by reducing the call on the volume of water at the termination of water plans)

Insisting that water users achieve a water efficiency dividend over time

Public investment in water use efficiency measures to reduce the overall call on the resource

Perhaps not surprisingly, most interest and public resources have been directed at the last policy choice (see, for instance [5]). After all, the political backlash from subsidising infrastructure investments in irrigated agriculture was unlikely to be too harsh, particularly given the deplorable state of knowledge about water affairs amongst the general public.

For instance, in September 2004 the Prime Minister announced the establishment of a $2 Billion Australian Government Water Fund with the lion's share ($1.6 Billion) dedicated to the Water Smart Australia Program which aimed primarily to "accelerate the uptake of smart technologies and practices in water use across Australia …[with most support] directed to practical on-the-ground projects" [7]. Similar enthusiasm for infrastructure refurbishment was evident in the $10 Billion *National Plan for Water Security* announced by Prime Minister Howard in January 2007. In this case almost $7 Billion was directed at reconfiguring irrigation through public investments in infrastructure.

The election of the Rudd government resulted in the release of *Water for the Future* in April 2008. In this instance there was to be $13 Billion expended over ten years. As with its predecessor, this policy foreshadowed that most expenditure would be directed at modernizing irrigation with $5.8 Billion assigned to "investment towards improving the efficiency and productivity of water use and management" [8].

In a departure from previous policies, *Water for the Future* dedicated a specific budget for the re-purchase of water to address over-allocation.

3 The contradiction

The basic premise for providing additional public investment to 'modernize' irrigation is that such investments will reduce 'waste' which can then be reassigned to deal with the over-allocation problem [9, 10]. In economic parlance, it might be argued that private irrigators will not undertake the optimal investment in infrastructure and public intervention is warranted on the grounds that the water purportedly 'saved' can then redeployed for the provision of a public good – say in the form of an enhanced riverine environment.

Clearly, proponents of the public failure doctrine might question the efficiency of this intervention. More specifically, several pieces of information would be required in order to make intervention with subsidized infrastructure efficiency-enhancing. First, the value of any environmental enhancement would need to be known lest the policy maker run the risk of over or under investment in environmental change. Arguably, this lack of information has bedeviled all forms of policy response in this context. Second, and most critical in this setting, is that there needs to be a reliable and verifiable means of articulating the water that is purportedly 'saved' as a result of 'modernizing' irrigation. Third, and in a related manner, mechanisms need to be in place to ensure that the water that is purportedly saved is then redirected to the provision of public goods and not appropriated by private interests.

Collectively, the last two information challenges have resulted in major public policy failures in numerous other settings [11], all in the name of 'water use efficiency'. More specifically, water use efficiency is itself a conceptually vexatious issue primarily because in an over-allocated or fully allocated catchment, water can seldom be 'saved' at a catchment scale [10, 12, 13]. Put simply, any calculation of water 'saved' on one farm (or one foodbowl) needs to also take account of the fact that others have existing claims to that which was previously spilled and perceived as 'going to waste' [14]. Several analyses are available to show that, in many cases, water use efficiency projects do little more than reallocate water in time and space, often away from the claimant with the weakest or least-well-defined rights [6, 15]. Moreover, the Productivity Commission recently notes that subsidising irrigation infrastructure is a poor use of taxpayers' funds, inconsistent with cost recovery principles agreed to under the NWI, can impede structural adjustment and is potentially inequitable to those who have already made investments in infrastructure [6].

An important upshot is that investing in such infrastructure potentially reduces the reliability of water, (i.e. reduces the probability of an irrigator receiving 100% of her allocation in a given year), particularly for interests downstream from the investments in modernization [16]. These events are made more acute by the fact that Australian jurisdictions operate under gross entitlement regimes, not net entitlements. Thus irrigators have an entitlement to exclusive access to water in each irrigation season, specified in volumetric terms or as a share of a specified consumptive pool. In simple terms, the supply of water is not increased for all by these projects, and for many users will actually be reduced. Thus, in order to maintain water-using activities at the same level, those outside the modernization project will invariably be forced to purchase additional water (i.e. to offset the decline in reliability). Alternatively, owners of these rights might chose to sell them, but it needs to be understood that they should now be of less value insomuch as their reliability has been degraded.

A second major ramification of investments in irrigation infrastructure relates to the direct impacts on the demand for water within the project area. Water is but one input used in irrigated agriculture. Thus, investments in irrigation infrastructure can be reasonably expected to impact on the marginal value product of water; each unit of water should be more productive by virtue of more timely application and greater control. After all, these arguments have been commonly invoked to make the initial call on the public purse [17]. Economists have long argued that the demand for inputs, like water, is a function of the marginal value product associated with that input. Thus, increases in marginal value product can be expected to result in increased demand for water.

Of interest here is the resulting impact on the availability and price of water in a market setting. The impact of irrigation modernization is twofold: reducing reliability of supply for those outside the project [9] forcing them to purchase more water rights or exit the industry while offering up their now lower reliability rights; increasing demand for water in agriculture within the project.

It follows that the price of water rights should rise [18].

This has significant implications for the operation of other policy approaches aimed at dealing with over-allocation.

Earlier it was noted that buyback of water rights is now a much-publicized part of the policy mix. Once demonized as the 'policy of last resort', the government purchase of water rights from willing sellers to secure increased environmental flows, is now openly acknowledged as a sensible means of "putting water back into rivers" [8]. More recently, attention has also been given to buyback as agencies strive to meet their obligations under the Living Murray program that sought to reassign 500 Gigalitres to the Murray River as a first step to achieving environmental balance.

Notwithstanding many reservations about the operational dimensions of current programs [19] buyback programs generally proclaim to be seeking "value for money" [20]. Surely, there are at least two critical elements where the simultaneous use of public funds to subsidize irrigation modernization is at odds with this criterion.

First, one component of value for money relates to the quantum of water that can be purchased from willing sellers for the least expenditure of public monies. Given that infrastructure projects raise the marginal product of water in agriculture, such projects actually reduce 'value for money' from buyback. More public funds will be required to secure a given quantum of water where buyback is accompanied by irrigation modernization.

Second, the intention of buyback is to improve the ecology of the Murray-Darling system by ensuring that there is sufficient water to meet environmental needs. To date, buyback has employed an expression of interest process seeking the involvement of willing sellers. It follows from the impacts of infrastructure on the demand for water that the most likely sellers in any water buyback program will be those outside the 'modernization zone'. Moreover, it has been argued that these water rights are now subject to lower reliability than might be historically expected as a result of the impacts on return flows due to 'modernization'. Thus, governments are not only forced to purchase rights at a higher price because of the public investment in irrigation infrastructure they are also likely to be purchasing rights subject to relatively low and declining reliability, again thanks to the investments in infrastructure. Buyback will yield lower reliability rights when accompanied by a modernization program.

4 Way forward

The contradiction described in the previous section represents a major conundrum for policy makers. Government leaders on both sides of politics have publicly espoused the benefits of irrigation modernization and to withdraw support carries non-trivial political risks. Some have advocated radical solutions, such as the acquisition of all water rights followed by an auction of a smaller quantum of rights back to those most willing to pay [21]. Such an approach is likely to be politically unacceptable and more pragmatic mechanisms are suggested here. These comprise both short and long term strategies.

The initial reforms to which all state jurisdictions agreed in 1994-5 required "that all future investment in new schemes or extensions to existing schemes be undertaken only after appraisal indicates that it is economically viable and ecologically sustainable" [22]. This condition remains implied within the National Water Initiative, although arguably it has not been given much weight in recent years. In a similar vein, much of the irrigation upgrades funded by the

Commonwealth was to be subject to 'due diligence' procedures before proceeding. A more vigorous application of these principles could potentially stall or even halt some of the more ambitious engineering escapades. This would, as a minimum, provide additional breathing space for buyback to be undertaken more effectively. It would also provide more opportunity to make public the benefits bestowed on many communities by buyback [23].

A major challenge will be the desire from some governments for Keynesian-style public infrastructure spending to deal with contemporary concerns about wider economic malaise. Nevertheless, there appears to be scope to deal with this through other expenditures without invoking the perverse effects associated with irrigation modernization.

Enthusiasm for projects in the 'water saving' genre stems from a naïve conceptualization of water resources and heroic assumptions about the capacity of engineering to deal with an 'inconvenient hydrology'. Both of these matters can be dealt with over time, especially by raising public consciousness of the costs that beset such endeavours. Research can play an important part here by empirically assessing the dual impacts of buyback and public infrastructure investments. The recent article by Lee and Acev [5] is a useful example of such work. However, in order to have a broader and more sustained impact it will be necessary to make research accessible to the voting (and taxpaying) public. The returns on expenditure directed in this area stand to be much higher than those that attend the present lavish expenditures on irrigation upgrades.

References

[1] Mercer, D., L. Christesen, and M. Buxton, Squandering the future--Climate change, policy failure and the water crisis in Australia. Futures 39, (2-3), 272-287 (2007).

[2] Musgrave, W., Historical "development" of water resources in Australia: Irrigation in the Murray-Darling Basin. Water Policy in Australia: The Impacts of Change and Uncertainty. L. Crase. Washington, RFF Press. 28-43 (2008).

[3] Davidson, B., Australia Wet or Dry? The Physical and Economic Limits to the Expansion of Agriculture. ed., ed. Vol. Melbourne, Melbourne University Press (1969).

[4] Quiggin, J., Environmental Economics and the Murray-Darling river system. Australian Journal of Agricultural and Resource Economics 45, (1), 67-94 (2001).

[5] Lee, L.Y. and T. Ancev, Two Decades of Murray-Darling Water Management: A River of Funding, a Trickle of Achievement. Agenda 16, (1), 5-23 (2009).

[6] Productivity Commission, *Market mechanisms for recovering water in the Murray-Darling Basin.* Draft Research Report, (2009).

[7] NWC, *National Water Commission Annual Report 2004-05*Canberra, National Water Commission.

[8] Wong, P., *'Water for the Future', A Speech to the 4th Annual Australian Water Summit*. Sydney Convention Centre., (2008).
[9] Molle, F. and H. Turral, Demand management in a basin perspective:Is the potential for water saving overestimated. International Water Demand Management Conference, Dead Sea, Jordan, (2004).
[10] Seckler, D., D. Molden, and R. Sakthivadivel, The concept of efficiency in water-reouurces management and policy. Water Productivity in Agriculture: Limits and Opportunities for Improvement. J. Kijne, R. Barker, and D. Molden. (2003).
[11] Perry, C., Efficient irrigation; inefficient communication; flawed recommendations. Irrigation and Drainage 56, 367-378 (2007).
[12] Molden, D. and R. Sakthivadivel, Water accounting to assess use and productivity of water. Water Resources Development 15, (1/2), 55-71 (1999).
[13] Perry, C., The IIMI paradigm: Definitions and Implications. Agricultural Water Management (1999).
[14] Gyles, O., *More water for irrigation and the environment? Some problems and prospects for worthwhile investments*. Connections. http://www.agrifood.info, AARES. (2003).
[15] Perry, C., Pricing savings, valuing losses and measuring costs: Do we really know how to talk about improved water management? the Management of Water Quality and Irrigation Technologies. J.D. Albiac, A. London, Earthscan. (2009).
[16] Sakthivadivel, R. and A. Chawla, *Artificial recharging of river water: An experiment in Madhya Ganga canal project, India*. Colombo, International Water Management Institute, (2002).
[17] DSE, *Victorian Water Accounts 2006-07*. Melbourne, (2008).
[18] Connell, D. and Q. Grafton, Planning for water security in the Murray-Darling Basin. Public Policy 3, (1), 67-86 (2008).
[19] Crase, L., S. O'Keefe, and B. Dollery, The Fluctuating Political Appeal of Water Engineering in Australia Water Alternatives 2, (3), 441-447 (2009).
[20] Breckwoldt, R., *Review of the 2007-08 Water Entitlement Purchases: Final Report*, Melbourne Hyder Consulting, (2008).
[21] Young, M. and J. McColl, Securing water: What is the best and fairest way to secure water for the environment? Droplet 18, (2009).
[22] CoAG, *Water Reform Framework*. (1994).
[23] Dixon, P., M. Rimmer, and G. Wittwer, *Modelling the AustralianGovernment's Buyback Scheme with a Dynamic Multi-Regional CGE Model,* General Papers No G-186. Monash University, Centre of Policy Studies and the Impact Project. (2009).

The cost of food safety due to animal by-product regulation in Spain: who pays for it?

A. Esturo[1], N. González[1], P. Greño[2], M. Martinez-Granado[2,3] & M. Saez de Buruaga[1]
[1]*AZTI-Tecnalia, Spain*
[2]*NAIDER, Spain*
[3]*Departamento de Fundamentos del Análisis Económico II, University of the Basque Country, Spain*

Abstract

This paper puts forward implications of the implementation of Regulation (EC) No. 1774/2002 in Spain, which regulates in an integrated manner all animal by-products that are not intended for human consumption with the maximum hygiene and environmental security. Farmers, slaughter houses, food industries, retailers, animal feed producers, fertilizer producers, transport and logistic businesses, renders, valorisation and waste management plants are activity sectors of great health, environment and economic importance and all of them play an important role in the implementation of the aforementioned regulation. This work analyses in detail the generation and management of the animal by-products in those different stages of the production chain of meat. The focus of the paper is on the transmission of the costs associated to the implementation of the regulation along the production chain. The results show an uneven distribution of the regulation costs, where the farmers face particularly high costs (which are, in general, subsidised by the Government). On the other hand, the transmission of costs to the final consumers depends on the type of animal we considered: for cattle and sheep, the lack of competitiveness in their markets also interferes in the costs transmission.

Keywords: food chain, modelling, pricing, animal by-products.

WIT Transactions on Ecology and the Environment, Vol 131,
www.witpress.com, ISSN 1743-3541 (on-line)
doi:10.2495/EEIA100071

1 Introduction

The European Parliament approved Regulation 1774/2002 as a consequence of the food health crisis, to regulate in an integrated manner all animal by-products with the maximum hygiene and environmental security.

The new regulation implied new management systems and higher costs. In Spain a specific Commission was created in 2003 to establish an Integrated National Plan for the management of the animal by-products and created different working groups for each step in the food chain. The results and conclusions were gathered in the White Book for animal by-products. One of these conclusions refers to the high costs for all the participating agents to fulfil the Regulation.

The application of the Regulation 1774/2002 implied that the animal by-products not intended for human consumption generated in every stage of the chain of value of meat and other animal products (from production to retailing) should be managed by authorised agents. In each stage of the value chain these agents should collect the animal by-products, charging a price that in principle has to be transmitted along the chain to the final consumers.

In this context it is important the identification of market distortions in the costs translation, difficulties in the gathering and transport of the by-products from generation locations to transforming and treatment plants, and the need to identify the possible valorisation of these by-products.

2 Objectives

The objectives of the study are:

- To elaborate a study on the economic aspects of the management of the animal by-products not intended for human consumption in the whole food chain to facilitate more transparency in the cost repercussion and reasonable translation of costs in the Spanish chain.
- To analyse the technologies available to obtain an added value from the animal by-products

3 Materials and methods

Farmers, slaughter houses, food industries, retailers, animal feed producers, fertilizer producers, transport and logistic businesses, renders, elimination and valorisation plants, and waste management plant, among others are activity sectors of great health, environment and economic importance. The analysis carried out evaluates the process, the costs, and the distribution of the agents through: (a) the identification of by-product generation processes, and determination of derived costs to implement the Regulation 1774/2002, (b) the characterization and quantification of the generated by-products, (c) the analysis of the economic aspects of the generation and management of the animal by-products and (d) the establishment of an efficient economic model for the animal by-product management.

The methodology used to carry out the study included the following phases:

Phase I: Diagnosis of the situation in Spain: actual infrastructures inventory, identification of the relevant markets, analysis of price and quantity evolution, market concentration and analysis of animal by-product waste management costs (field work).

Phase II: Comparative analysis: comparison of the experiences in other European countries applicable in Spain.

Phase III: Theoretical economic frame of reference: the results of the previous phases allowed designing a theoretic economic frame *ad-hoc,* based in traditional supply-demand models. The main elements of the designed models are: price formation along the value chain of different animal products; transfer of the cost increments among agents; analysis of the inefficiencies and completion problems on the cost (price) transmission along the value chain of the animal products.

Phase IV: Estimation and results of the economic model: econometric estimation of price transmission along the different agents involved in the value chain of different animals (beef, sheep, pork, and chicken). This information allowed to evaluate the absence of competition in different stages of the value chain of the animals considered and therefore to throw conclusions about the final transmission of the by-product new management costs to the final consumer.

Phase V: Animal by-product valorisation: new management solutions and valorisation options are identified for the Spanish context.

4 Results

According to the data provided by the Spanish National Commission of the SANDACH (www.sandach.com.es) in 2009, 623 enterprises had the license to manipulate animal by-products of different categories according to Regulation 1774/2002 (see the SANCO/10149/2006 report) [1]. The typology of the processors is 15,5% process plants, 38,7% specific users, 17% technical plants, 13,7% intermediate plants, 4,2% storage plants, 4,4% plants of incineration or co-incineration, and 3,4% plants of compost. The process plants process most of the 2.000.000 tons generated in Spain. It is significant that no approved biogas plants are in the list. Most of the plants are located in the North of Spain (Castilla-León, Galicia, and Catalunya), where most of the bovine, ovine and pig production is located. Another striking fact is that only two plants are producing fish oils from extractive fishery wastes, but no aquaculture wastes are being used. Comparing the animal by-product plants in other European countries (Great Britain, Germany, France and Denmark), it is outstanding the number of biogas plants in Germany, whereas in Great Britain, a high number of incineration and co-incineration plants are registered. This could be due to the fact that the Bovine Spongiform Encephalopathy (BSE) started in the U.K. Spain has a very high number of specific users, processing and technical plants, when compared to other countries.

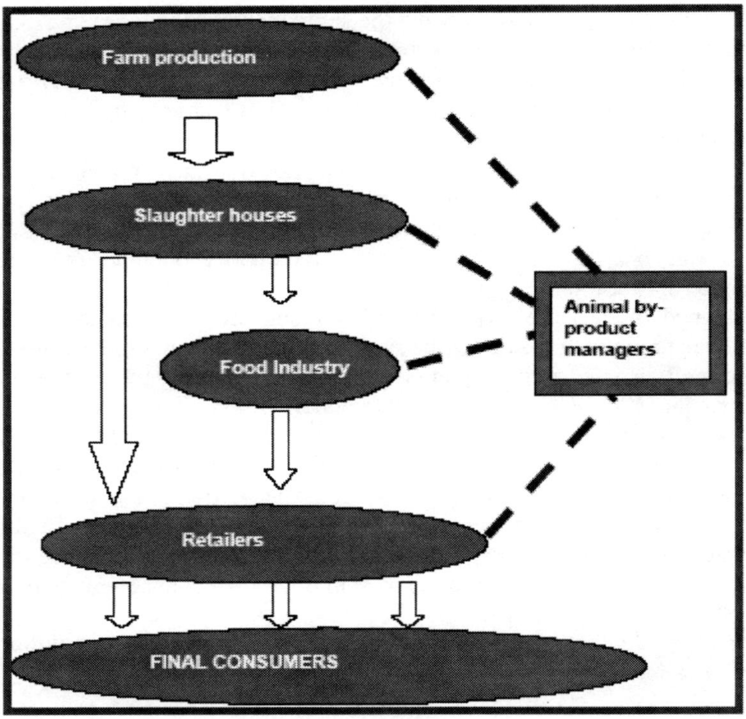

Figure 1: Price generation in the food chain.

The agents identified in the Spanish animal by-product management chain go from farmers to consumers as shown in the figure 1.

The animal by-product generators are farmers, slaughter houses, food industries, and retailers. The processed materials are gathered and eliminated, or transformed or valorised by by-product managers. Most of the by-products are produced in the meat industry, although part is also produced in companies related to fish commercialisation and transformation industries.

The first step in the chain is the farm production. The data on production, stocks and costs are relatively simple to obtain through the Spanish Ministry of Agriculture, Food and Fisheries (now Ministry of Environment, Rural and Marine Affairs, MARM). The main production includes bovine, ovine, pig and poultry production. There is a considerable atomization of the production with an average size of the production sites of 43 heads for bovine, 231 heads for ovine, 197 for pigs and 886 for poultry. Pig production accounts for more than 40% of the total animal production considered (bovine, ovine, pig, and poultry), being Spain the second producer in Europe after Germany. (Source: own elaboration with data from the "Spanish Survey on the structure of the agricultural exploitations", National Institute of Statistics-INE, 2005.)

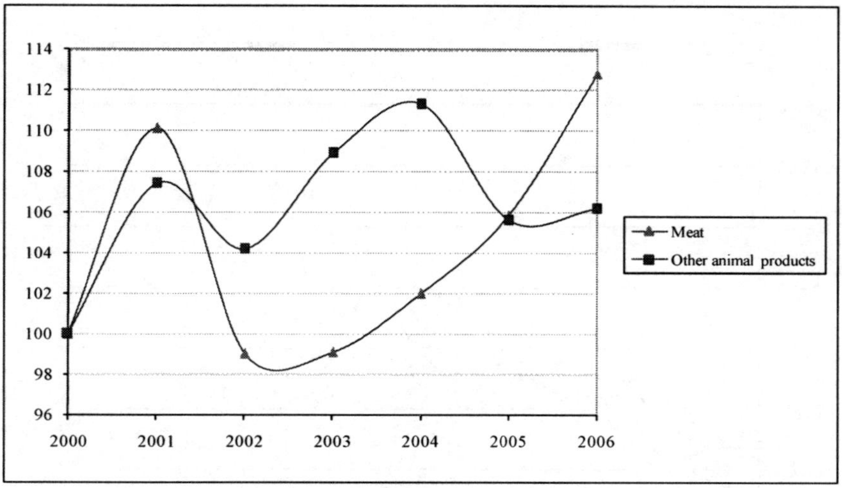

Figure 2: Prices index for meat and animal products. Index year 2000=100 (source: MARM [3]).

Prices on the meat market are quite cyclical. Figure 2 presents the evolution of the index of prices obtained by farmers for meat (beef, sheep, chicken, pork, goat, and rabbit) and animal products (milk, eggs and wool).

The prices for animal products are much more stable than prices for meat. A high fall can be appreciated between 2001 and 2002, probably due to the decrease in the demand of bovine meat in those years. This fact is important in terms of the implications of assuming the higher costs derived from the application of the Regulation.

Slaughter houses and meat industries are the second step in the value chain. An increase in pig and poultry slaughtering is observed, being in 2005 about 58% of the slaughtered animals pigs and 23,5% poultry. As far as the structure of this step is concerned, it must be said that it is not easy to find updated data for the sector although an important atomisation of the sector is observed, with many SME (less than 11 workers). A tendency for vertical integration of the industry with the production stage is observed, mainly in the poultry and pig production (see Langreo Navarro [2] and MARM [3]). This vertical integration is important in the cost transfer: the more integrated the different steps in the value chain are the easier the cost transmission will be, as it happens with poultry and pig. The prices for meat in the slaughter houses have been increasing steadily since 2001.

Meat industry is one of the most important sub-sectors of the Spanish food industry, generating more than 16.000 million Euros in 2006. There is an important leader in this sector, and other 40 medium-big size companies with a high number of small traditional businesses. Only 23% of the companies have more than 20 employees [2].

Retailers are the last step in the chain, supplying meat to the consumers (wholesalers are not important in this sector).In 2005, 2818 tons meat and 1572

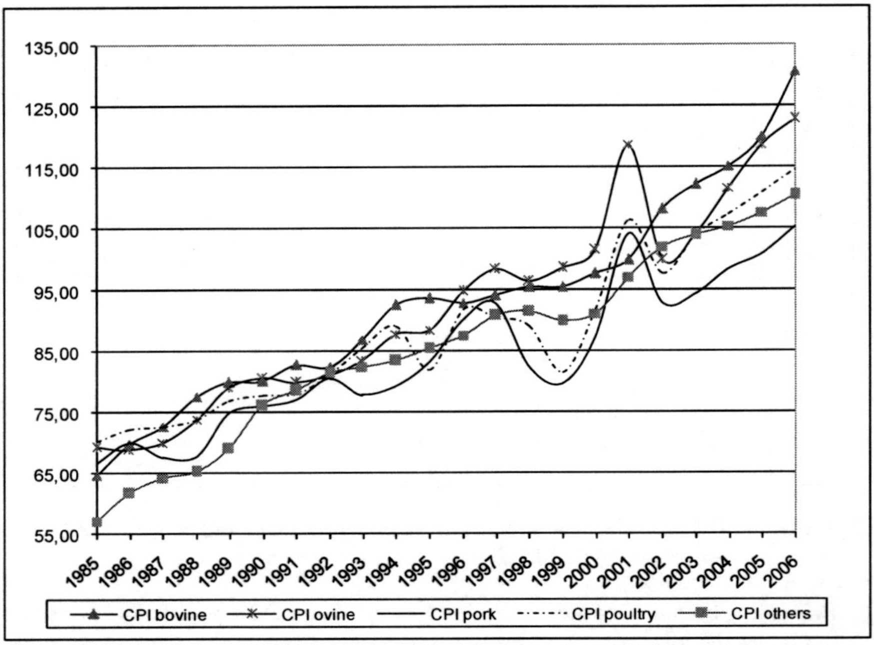

Figure 3: Consumer price index for meat consumption (source: Spanish National Statistics Institute [13]. Base year 2001=100).

tons fish products were consumed (see MERCASA [4]). Fresh meat and fish is mainly bought in traditional markets and supermarkets, and not so much in hypermarkets. The price for meat products consumption has grown steady in the last years as shown in the figure 3.

4.1 Costs of by-product management

Little information is available about the real amount of by-products generated and the costs implied by the application of the Regulation. Field work was done to collect the needed data, through personal and telephone interviews. Specific questionnaires were designed for each type of agent (farmers, slaughter houses, food industries, retailers and by-product managers) asking questions about volumes, typology of by-products, costs and inversions required to implement the Regulation (CE) 1774/2002. Finally 1012 agents were identified, where 294 were selected. 54 agents were personally interviewed and 250 agents were interviewed on the phone. This numbers include information from some firms obtained from the National Agrarian Insurance (ENESA) en the retailer association (ASEDAS).

The main results from the interviews are the following: In general, the application of the Regulation 1774/2002 has implied a cost to all agents related with the production and management of animal by-products. This was the

Table 1: Number of interviews.

Sector	Interview number
Farmers	67
Slaughter houses	59
Food Industry	55
Retailers	60
Managers	38
Incineration	15
TOTAL	294

Table 2: Estimated costs of insurance for meat as a percentage of the farm price.

	Total cost (€) per 100 kg (1)	Farm price (€) per 100kg (average) (2)	% over obtained price (3)=((1)/(2))x100
Bovine	6,4	127,5	5,1
Ovine	4,3	268,5	1,6
Swine	1,7	108,0	1,5
Poultry	0,9	89,0	1,0

(1): Total cost per 100 kg was calculated using as reference the average weights as described in the "Plan Integral de SANDACH".
(2): farm price is calculated considering the slaughtered volumes of different types of animals. Data from MARM [11, 12].

answer of 53% of the farm producers, 80% of the slaughter houses, 52% of the retailers, 45% of the food industry producers, 66% of the by-product processors and 78% of the (co-)incineration plants and dumps. Often by-products of category 1 and 2 are indistinctly treated.

The increase in the costs has been double: on the one side the investments in infrastructures and equipments to treat the by-products have been high, specially for some agents (slaughter houses, incineration and co-incineration plants, dumps, and by-product processors); on the other side, the marginal costs of processing in each stage the by-products has also increase. However this cost differs for the different agents implied.

For farmers, the cost has been mainly due to the management of the category 1 of by-products, which implies the collection and destruction of dead animals in the exploitation. This operation is usually paid by an agrarian insurance, heavily subsidize by the local or central governments. The costs are presented in table 2. As it can be seen the costs of retiring dead animals imply a 5.1% of the price received by the bovine farmers, and between 1 and 1,6% for the rest of the farmers.

However, it has to be mentioned that bovine farmers have an extra cost that slaughter houses apply when slaughtering the animals to cover the destruction of the by-products, which is done according to the legislation Orden APA/67/2002,

Table 3: Cost of management of by-products in the different sectors (€/kg of by-product; it does not include the retrieval of dead animals from exploitations).

		Cat. 1	Cat. 2	Cat. 3
Farms of bovine	(to slaughter houses)	0,48-0,52		
Slaughter houses		0,32	0,17	0,12
Industry	Average	0,22	0,06	0,06
	Meat	0,22	0,06	0,09
	Fish		0,12	0,02
Meat retailers	Average			0,28
	Central Markets			0,14
	Hypermarkets			0,15
	Supermarkets			0,25
	Traditional shops			0,23

which establishes a control systems for the food by-products (see table 3). The idea of the regulation was to give transparency in the by-product generation and the related costs. However, it seems that slaughter houses apply the costs to the farmers, instead of to the retailers. Not only that, the price charged to the farmers do not correspond to the one charged by the by-product managers, therefore they use this activity as another lucrative aspect and they have certain market power (they charge more than the marginal cost they pay the by-product managers). Farmers pay twice the cost of by-product elimination (through the agrarian insurance and when selling the animals). In a competitive market the firs cost could be passed on to the second chain, but the second cost no.

For slaughter houses, table 3 shows the average cost for the disposal of the different categories of by products. It is important to mention the high dispersion of costs across regions, being more expensive to manage the by-products in the North of Spain.

With respect to the food industry, the meat sector is the one with higher costs. In some cases, the slaughters take care of the costs of by-product processing. It has to be mentioned that in the case of fish transformers there is a special difficulty in the identification for the by-product manager.

Small retailers (traditional shops) declare in general not to incur in any costs (64%). Central markets and hypermarkets obtain better prices than smaller shops. They tend to pay higher prices than the food industry or the slaughter houses.

Finally, it is important to mention that although the by-products processing plants have also incur in important costs to adapt their installations, they have also received higher subsidies than in other sectors. Additionally there is evidence that they have some market power to set prices, as shown by the geographical dispersion of prices charged and the comparison with the cost of transformation or disposal of the by-products.

4.2 Economic frame and cost transmission

Once established the increase in costs in each stage of the value chain of meat and fish, here we concentrate in the transmission of costs along the meat value chain, the one that concentrates the higher costs. If the markets work in perfect competition, the increase in the marginal cost of production of meat would imply an increase in the sale price in each stage equivalent. Therefore the higher costs implied by the Regulation will be to some extend passed to the final consumers as intended and to some extent assumed by the producers in each stage of the value chain, depending on the supply and demand curvatures. See Unnevehr et al. [5] or Radwan et al. [6] for evidence in these curvatures.

However, if the market is not perfectly competitive, some of the firms participating in some of the stages of the value chain can avoid their portion of the costs by imposing prices to the following or previous stages. They would use their market power (as sellers or buyers) to avoid these costs. In this section we analyse whether this is the case. For doing so, we estimate an econometric model in which the prices charged by retailers (and paid for the consumers) are a function of the prices charged by the farm producers (there is no information about the prices charged by slaughter houses). We follow Tomek and Robinson [7] and Jumah [8]. The equation estimated is:

$$P_{ft} = c^* + a^* P_{gt} + \varepsilon_t$$

where P_f is the price paid charged by the retailer and P_g is the price charged by the farm producer. The test of market power is done through the test of the null hypothesis that $a^*=1$. The estimation follows the Engle and Granger [9] two stage procedure that accounts for non stationary of the series and uses a Vector Error Correction (VEC) econometric specification.

The main results of the estimation are presented in table 4. No market power is apparent in the pork and chicken markets, but its presence cannot be rejected in the case of the beef and sheep markets. The results are compatible with the existence of market power from the slaughter houses already indicated in the

Table 4: Results from the econometric analysis: long run relationship.

Long-run relationship								
	Beef		Pork		Sheep		Chicken	
	Coeff.	t-stat.	Coeff.	t-stat.	Coeff.	t-stat.	Coeff.	t-stat.
P_t^g	3.87*	6.62	0.77*	5.38	7.77*	-6.14	0.90*	6.55
Constant	2.10*	1.82	0.30*	17.44	20.46		1.56*	11.37
Draft	YES		YES		NO		NO	
Time Dummy	YES		YES		YES		NO	
Seasonal Dummies	NO		YES		YES		NO	

* Significant at 5%

previous section with respect to the bovine sector, although some market power from retailing might be also present, especially if the market concentration of the sector in recent years is taken into account.

4.3 By-product management and valorisation

By-products are traditionally transformed to eliminate the water and extract the fats. The products of this transformation are meals or animal proteins and fats used as feed, or in chemical industries for soap production.

There are other transforming alternatives already used in Europe, which can be classified according to their energetic component in three different groups: low, medium and high energy (see Woodgate [10]).

4.3.1 Low energy

The management alternatives for animal by-products range from burials, dumps, compost windrow production, anaerobic digestion (biogas production) and liquefaction.

4.3.2 Medium energy

In this category process for pet food production, alkaline hydrolysis, high pressure hydrolysis, biogas production by pressure hydrolysis, biodiesel production and Brookes gasification process can be found.

4.3.3 High energy

The incineration is an oxidation at high temperature to transform the product into gases and ashes. There are different types of incinerations: Bubbling Fluidised Bed, Circulating Fluidised Bed, Rotary incinerator oven, and continuous incineration.

After analysing the environmental effects of each of the alternatives, as well as the induced effect, the applicability and the economical aspects, it is difficult to compare the different technologies, and therefore it is not simple to make recommendations. Considering environmental criteria (air emissions, effluents to waters, effluents to soil, energetic resources and CO2 balance) and the sustainability criteria (waste hierarchy, regulations, capacity, society and environmental troubles) the best options are: (a) the transformation into meals and fats for animal feed, (b) transformation into meals as fertilizers, (c) transformation into meals to be used as fuel in the plant and (d) incineration with energetic recovery. The worst option is the disposal in a controlled dump (with or without energetic recovery).

5 Conclusion

Application of Regulation 1774/2002 has incremented the costs in the Spanish meat production chain and to a lesser extent fish production chain. The cost increase in most of the cases implies acquisition of infrastructures and equipments, and an increase of the marginal costs per production unit for the animal waste generators. It is observed that animal by-products of category 1 and 2 are managed with no distinction. This is an identified aspect for improvement. According to the observed data it seems that by-product renders and slaughter houses have certain market power at least in the bovine and ovine markets. Renders generate higher prices than expected, and slaughters pay less than expected in a competitive market. Nevertheless the econometric analysis (price of meat for farmers and price of meat for consumers), partially confirms the hypothesis: while the pig and poultry markets are competitive, the bovine and ovine market, have difficulties to transmit the prices and cost of application of the Regulation.

Acknowledgements

This paper has been written as part of the study “Estudio sobre la distribución de costes de gestión del los SANDACH a lo largo de toda la cadena” made for the Subdireción General de Mercados Exteriores y Producciones Porcina, Avícola y Otras, del Ministerio de Medio Ambiente y Medio Rural y Marino (MARM). We thank the MARM for allowing us to publish the results.

References

[1] SANCO/10149/2006. “Technical specifications in relation to the master list of lists and the list of approved plants handling animal by-products”, European Commission.

[2] Langreo Navarro, Alicia. “La industria cárnica en España”. Distribución y Consumo, 5, May-June 2006.

[3] MARM (Ministry of Environment and Rural and Marine Affairs), “Sector cárnico español”, 2001.

[4] MERCASA, “Alimentación en España 2007: producción, industria, distribución y consumo”, 2007.

[5] Unnevehr, L.J., M.I. Gómez, and P. Garcia. “The Incidence of Producer Welfare Losses from Food Safety Regulation in the Meat Industry”, Review of Agricultural Economics, 20 (1), pp.186-201, 1998.

[6] Radwan, A., Gil, JM., Ben Kaabia, M., Serra, T., “Modelling the impact of food safety information on meat demand in Spain”. 107th European Association of Agricultural Economists Seminar, Sevilla, 2008.

[7] Tomek, W.G., y Robinson, K.L., “Agricultural Product prices”, Cornell University Press, Ithaca, 4rd edition, 2003.

[8] Jumah, A., “The long run, market power and retail pricing”, Empirical Economics, 29, pp. 605-620, 2004.

[9] Engle, R.F., y Granger, C. W. J., “Cointegration and error correction: representation, estimation and testing”, Econometrica, 51, pp. 277-301, 1987.

[10] Woodgate S., “What would a world without rendering look like?” in Essential rendering: all about the animal by-products industry, 2007.

[11] MARM (Ministry of Environment and Rural and Marine Affairs), "Survey about farm prices for agriculture and stockbreeding", 2005.
[12] MARM (Ministry of Environment and Rural and Marine Affairs), "Survey of cattle slaughter", 2005.
[13] Spanish National Statitistic Institute (INE), www.ine.es.

Section 2
Cost benefits analysis

Cost-benefit risk of renewable energy

K.-J. Hsu
Department of Construction Technology and Facility Management, Leader University, Taiwan

Abstract

State-of-the-art economic analyses of renewable energy technologies (RETs) include using portfolio theory in renewable energy policy planning and using Monte-Carlo simulation to attain the risk profile of the technologies. Most economic analyses now co-list the risk index and GWP index as results of research/policy planning reports. After examining the variants of economic models for RETs, the risk dimensions of traditional energies and RETs were compared. The asymmetry risks allocated between traditional energy and RETs were shown. Combining these asymmetry risk allocations, a project-based RETs financial feasibility model which simultaneously integrates cost and risk information with respect to renewable and traditional energy technologies, was developed. Results of the analysis showed that the risk premium of traditional energies should include the effects of the escalation and volatility of fuel prices. On a national/regional level, the subsidy of fossil fuels for traditional energy production, which is embedded in the economic development policy, should be recovered and the degree to which investment of RETs improved the efficiency of cost-risk portfolio should be measured. On a project-based level, the effect of fuel price escalation and the saving of cost of risk-mitigation should be included.

Keywords: energy economics, project-based analysis, economic feasibility, risk evaluation, energy risk.

1 Introduction

When analyzing the economic feasibility of a renewable energy technology, both negative externality and price risk of traditional energy technologies (TETs) are major concerns, as well as the difficulty regarding integration into the evaluation model. Hsu et al. [1] used life cycle cost analysis and a saving-to-investment ratio to analyze the empirical data of a residential BIPV

WIT Transactions on Ecology and the Environment, Vol 131, © 2010 WIT Press
www.witpress.com, ISSN 1743-3541 (on-line)
doi:10.2495/EEIA100081

project located at Kao-Hsiung [1]. The results showed that compared to the low electricity price in the southern part of Taiwan, it was still not economically feasible for a home owner-occupant to embrace renewable energy, even with a government subsidy. Hsu et al. (2008) used discounted payback year and two-way sensitivity analysis to analyze the future feasible domain of BIPV based on empirical data of a residential BIPV project, but did little to address the implication of the risk-mitigation with respect to the price volatility of fossil fuel [2].

State-of-the-art technology for dealing with such a problem includes portfolio theory and Monte-Carlo simulation. Some researchers and renewable energy policy-makers used the portfolio theory to find the optimal renewable proportion which can decrease the risk of energy combinations at the national/regional level, because the cost risk of renewable energy always exhibits a low correlation to traditional energy technology. For example, some EU reports used Energy technology portfolios to address energy security and diversity [3–5]: Airtricity use of wind generators to enhance Scotland's energy diversity and security [6], along with California's 33% renewable portfolio standard goal [7].

Other approaches used the Monte-Carlo simulation results and co-listed the risk dimension for reference in the analysis, like: net present value (NPV), internal rate of return (IRR), life-cycle cost (LCC), and levelized cost of energy (LCOE/LOE). Only a few analyses now co-list the risk index and global warming potential (GWP) index as results of research/policy planning reports [13]. Some of them use the market transaction value of GWP as a cost component of the energy technologies. But determining how risk attributes can be integrated within an economic analysis framework still needs be developed.

In the following section, the paper first examines the variants of economic analysis for RETs. Then the risk dimensions of RETs and traditional energy technologies are discussed in Section 3. A project-based economic feasibility model for RETs under uncertainty is developed in Section 4. Finally, a discussion is offered and conclusions are drawn.

2 Variants of economic model for RETs

Different approaches were used in the economic analysis of renewable energy, e.g. NPV, IRR, LCC, and LCOE. When compared to an alternative with a lower initial cost but a higher future cost, the pervasive tool used in engineering is the LCC. The LCC practice establishes a procedure for evaluating the life-cycle costs of a system and comparing the LCCs of alternative systems that satisfy the same functional requirements. The method entails computing the LCC for alternative building designs or system specifications having the same purpose, and then comparing them to determine which has the lowest LCC over the study period [9].

Since RETs are always used as substitutes for current energy technologies, energy saving and environmental impact are treated as major benefits. RETs always have higher initial costs and lower future costs (operating, maintenance, repair, or replacement costs); under these assumptions, it seems suitable to use in an economic analysis of renewable energy and an energy efficiency scheme. The present value of life-cycle cost ($PVLCC_e$) for an energy technology can be presented as follows:

$$PVLCC_e = IC_e + PVM_e + PVR_e + PVF_e - PVS_e \tag{1}$$

where IC is the present value of the initial investment cost; PVM, operation and maintenance; PVR, facility replacement cost; PVF, fuel price expenditure; PVS, energy saving; and subscription $e = te$ (traditional energy), re (renewable energy). Under the same theoretical concept, LCOE was developed to compare different energy production technologies. Most energy policy research organizations, like IEA, USDOE, or textbook used LCC or LCOE to simplify the analysis of RETs, e.g. in [1, 8, 10, 11, 14, 15].

Theoretically, the decision process in selecting between the alternatives should include consideration not only of the comparative LCCs/LCOEs of competing systems, but also the risk exposure of each alternative relative to the investor's tolerance for risk, any unquantifiable aspects attributable to the alternatives, and the availability of funding as well as other cash-flow constraints. Sensitivity analysis and probability analysis are the two major tools used in determining risk and uncertainty. In case of RETs evaluation, we always include the different price risk and externality scenarios in all the alternatives. Because $PVLCC_{te}$ and $PVLCC_{re}$ are evaluated separately, we list the resulting PVLCC and risk dimension in a descriptive table. IEA/OECD (2005) listed both the environmental externality index (e.g. GWP) and the risk index (e.g. standard deviation) of each energy technology, as supplementary to the comparisons [6]; however, the report did not provide full information regarding the impact of fuel price risk.

Recently, at the level of national/regional energy policy, more and more researches have used the portfolio theory to deal with the cost-risk profile of energy. The portfolio theory approach used mean-variance analysis to find the efficient frontier of different energy technology combinations, thus deciding the optimal proportion of renewable energy technologies. For example, Awerbuch and Berger (2003) used the portfolio theory to add renewable energy to the EU traditional generation mix [5]. Awerbuch (2005) used the same approach in "Wind Provides Competitive Advantage for Scotland" [6]. Bates White, LLC, (2007) used the same approach in *Achieving California's Generation Mix to 2020*: California's 33% Renewable Portfolio Standard Goal [7].

3 Risk dimensions of TETs vs. RETs

When evaluating the financial feasibility of a PV system, this paper refers to the capital investment in a photovoltaic electric generated system in which the initial capital cost can induce streams of cash inflows without fuel input and less environmental negative externality (e.g. GWP) at a minimum maintenance and

Table 1: Comparisons of risk characteristics of TETs vs. RETs.

	Input/ Cost components	Output / Benefit and Externality
RETs	1. More expensive initial capital investment compared to traditional fuels 2. Zero fuel price 3. Quantity of electric power generated fluctuates according to weather variations 4. Little maintenance and low operation fee across facility life cycle	1. Electricity power 2. Little GHG emission/GWP
TETs	1. Sunken cost of former capital investment 2. Fuel price with high escalation rate and volatility rate ***National/regional scale:*** 3. Uneven fuel distribution affecting energy security of the country 4. Global fuel potential exhausted	1. Electricity power 2. Large GHG emission/GWP

operation fee. Normally, the cash inflow of the PV investment is an opportunity to benefit from the substitution of a traditional electrical system generated from a primary energy material like fossil fuel. Compared to the new renewable technologies, one of the most prominent characteristics of the traditional energy systems is a sunken cost with escalating and fluctuating fuel prices, followed by an environmentally negative externality.

Table 1 presents a summary of the risk characteristics of cost components in regard to renewable and traditional energy technology. Focusing on PV electric power generated from photovoltaic system, due to the uneven distribution of such primary energy material (e.g. coal, oil and natural gas) across the continent, they are quite concentrated in some specific areas. This not only results in complicated geo-political relationships, but also results in energy security problems in most countries in the world and affecting the vulnerability of the economic development of most countries.

Based on the primary comparison among renewable energy and traditional energy in the above paragraph, the major differences between renewable and traditional energy include ***Risk and Externality.*** In terms of ***Externality***, the negative externalities accompanying traditional electric power production process are always on a global scale. Although it is hard to include externality cost completely in a practical way, it is possible to use the GWP market transaction price as proxy and compute it into the evaluation. In terms of ***Risk***, an examination of the effects of different risk properties regarding renewable and traditional energy technology reveals that RETs are always viewed as an intermittent energy source; as a result, the stochastic production quantity always involves variability. The risk of traditional energy technologies is mainly based on fuel price escalation and fluctuations along the whole life span.

Variability of RETs

Sinden [17] used UK empirical data to show that onshore wind speed correlations rapidly decrease as distance between wind farms increase [17]. Thus, the variability of wind energy in combination with dispersed location will significantly lower the electric generation risk. Empirical data in Germany also showed high wind in the winter and more sun in the summer [3]. The inverse correlation of seasonal capacity factors (actual power output divided by maximum potential output) of wind and PV can be complementary via careful renewable planning; this showed that the variability of renewable issues can be transformed via management strategies, technical system integration, and planning processes.

Fuel price risk

Market risk of the traditional energy production includes fuel price escalation and fluctuations. There are two kinds of approach that deal with market risk: national/regional policy level and project-based level. As we discussed in the previous section, national/regional policy uses the portfolio theory to determine the optimal level of RETs. At a project-based level, if we hope to combine the market and production uncertainty in the framework of the economic analysis of RETs, we will face the problem of computing energy-saving functions. How can the risky information on traditional energy technology be integrated into an energy-saving function? Some transformational method needs be developed in advance.

By examining the risk properties within different energy technologies, we find the asymmetry risk between renewable and traditional energy. This will definitely affect the result of project-based economic analysis. These biases will hinder research results. For example, the fuel price risk in a traditional energy technology will result in higher risk-premium cost and cost transfer to the end user. Also, whenever analyzing the renewable technologies, this risk from the counter-side alternative energy cannot be assigned to an energy-saving function. Whenever the asymmetry risk is misallocated, biases will follow. Without proper treatments of the model specification, these asymmetry risk properties will affect the result of a project-based economic analysis; thus, biases will definitely hinder research results.

4 Project-based model for RETs under uncertainty

If the benefits and costs can be completely discounted, net present value of a RET investment can be evaluated by inverting the sign of the right hand side in

Eq. (1). By comparing the risk attributes of additional investment of renewable energy respect to each items of traditional energy, the cost-benefits of renewable energy is developed, as shown in Table 2.

Table 2: The cost-benefit of choice energy technologies.

	Items (present value)	**RETs**	**TETs**
Cost	Initial investment cost	C_{re}	C_{te}
	Operation and maintenance	M_{re}	M_{te}
	Facility repair or replacement cost	RP_{re}	RP_{te}
	Fuel price expenditure	*0*	F_{te}
	Cost of environment externality/GWP	GWP_{re} (small)	GWP_{te} (large)
Benefit	Electricity power	Electricity power	Electricity power
NPV	Net present value	PW_{re}	PW_{te}

Let PW_{te} denote the present value of traditional energy technology. C_{te} denotes the present value of the sunken value of the capital investment, M_{te} denotes the present value of the annual operation and maintenance costs; RP_{te} denotes the present value of the repair or replacement fee of the energy facility; F_{te} denotes the cost of the fuel and GWP_{te} denotes the cost of *GHG* emissions which were summarized as an index of global warming potential (*GWP*). Let A_{te-0} denote the initial bill of the energy user at time 0. Without considering the escalation effect of energy prices, the net present value of unit investment in traditional energy with life span, *n,* can be rewritten as follows:

$$PW_{te} = -C_{te} - M_{te} - RP_{te} - F_{te} - GWP_{te} + A_{te\text{-}0}\frac{(1+i)^n - 1}{i(1+i)^n} \tag{2}$$

The first five terms on the right hand side of Eq. (2) are a summation of the present value of the cost components; the last term is the present value of the cash inflow from electric bills. Normally, if the net present value is positive, the evaluated project is feasible. Supposing the energy market is completely competitive and the normal profit of the investment is zero, then $PV(B_{te})=PV(C_{te})$ can be set without loss of reality. The equivalent computed electric price (A_0^*) of the traditional energy-k (including the cost of environmental externality and risk premium of the fuel escalation rate and volatility) at time zero thus can be obtained; therefore A_0^* can be interpreted as the risk-adjusted levelized cost of traditional energy technology.

The present value of energy savings for renewable energy investment in Eq. (1) will show the fuel market price risk of traditional energy technologies (PVS_{te}). However, due to the asymmetry risk between traditional energy and RETs, the stochastic property of fuel price risk should not be directly specified in the investment value function of RETs. The risk premium of risk-mitigation of RETs should be counted as part of energy saving, thus raising the normal traditional electric market price. The traditional electricity price function with price fuel risk includes a high escalation rate (*e*) and high volatility (*v*), thus raising the opportunity cost of risky capital investment.

Thus the present value function of a RET, like BIPV can be written as follows,

$$PW_{re} = -C_{re} - M_{re} - RP_{re} - GWP_{re} + A_0^* \frac{(1+i)^n - 1}{i(1+i)^n} \tag{3}$$

In Eq. (3) the cost of the fuel F_{re} is zero. The present value of operation and maintenance cost M_{re}, the repair or replacement fee of the energy facility RP_{re}, and the cost of environmental externality GWP_{re} of the renewable energy, like BIPV, is rather small. But the risk-premium of the fuel price risk of the traditional energy is now positive for renewable energy, which can be considered in imputed equivalent electric price A_0^*. At the national/regional level, the subsidy of fossil fuel for traditional energy production embedded in economic developed policy should be recovered, ΔA_1.

The effect of fuel price escalation

Considering that the rate of energy price escalation trend (*e*) is always faster than the rate of general price inflation, the final term of Eq. (3) can be replaced as $A_0 \sum_{j=1}^{n} \left(\frac{1+e}{1+i} \right)^j$, which denotes the present value of energy saving for TETs.

The adjusted present value factor of energy saving can then be derived as follows,

$$\sum_{j=1}^{n} \left(\frac{1+e}{1+i} \right)^i = \begin{cases} \dfrac{(f^n - 1)f}{f-1} & , if\ f \neq 1. \\ nR_0 & , if\ f = 1. \end{cases} \tag{4}$$

in which $f = \dfrac{1+e}{1+i}$; and *i* is the discount rate.

The effect of fuel price volatility

The risk premium of the computed equivalent electricity price A_0^* should include both the effect of fuel price escalation and volatility in the market. The benefit from improving the efficiency frontier for the cost-risk portfolio of social welfare (ΔA_2) with respect to the specific RETs investment and energy security (ΔA_3) should be measured and added to the benefit terms of RETs. On a project-based level of project feasibility evaluation, the opportunity cost of risk-mitigation for traditional energy production with respect to fossil fuels ΔA_4 should be estimated and added to the opportunity benefit terms of RETs.

On a national/regional level, the subsidy of fossil fuels for traditional energy production, which is embedded in the economic development policy, should be recovered and the investment of RETs improved by the efficient frontier of cost-risk portfolio as well as the energy security of the society should be measured as: $\Delta A_1 + \Delta A_2 + \Delta A_3$. On a project-based level, the effect of fuel price escalation and the saving of opportunity cost of risk-mitigation should be included: ΔA_4. Based on the computed equivalent electricity price of the traditional energies, which includes the adjusted factor of price escalation of energy-saving, the relevant risk-premiums for Eq. (3) can be summarized and expressed as follows:

$$\Delta A = \Delta A_1 + \Delta A_2 + \Delta A_3 + \Delta A_4 \tag{5}$$

5 Conclusion

By examining different risk dimensions of energy technologies, the asymmetry risks allocated between and TETs and RETs were shown. Combining the attributes of the asymmetry risk, a project-based economic feasibility model of RETs was developed. Such a model can simultaneously integrate all of the cost and risk information of RETs and TETs into one. This showed that the parameters of asymmetry risk of the two kinds of energy technologies and the environmental externalities could be adjusted. The risk premium of traditional energies should include the effects of escalation and volatility of fuel price. On a national/regional level, the subsidy of fossil fuel for traditional energy production which is embedded in the economic development policy should be recovered and the investment of RETs improving the efficient frontier of cost-risk portfolio should be measured. On a project-based level, the effect of fuel price escalation and the saving of opportunity cost of risk-mitigation should also be included.

Acknowledgement

The authors would like to thank the National Science Council of the Republic of China for financially supporting this research under Contract No. NSC 97-2221-E-426-012-MY2.

References

[1] Hsu, K-J, Lai, C-M, Chiang, C-M, The Economic Evaluation of Building-integrated Photovoltaic Systems, *Proceedings, Annual Conference - Canadian Society for Civil Engineering, v 1, Annual Conference of the Canadian Society for Civil Engineering 2007: Where the Road Ends, Ingenuity Begins*, pp. 390-397, 2007.

[2] Hsu, K-J, A projection of BIPV systems in Taiwan: costs versus benefits. *Environmental Economics and Investment Assessment II*, WIT Transactions on Ecology and the Environment, Vol 108, pp. 161-170, WIT Press, 2008.

[3] IEA, *Empowering Variable Renewables - Options for Flexible Electricity Systems*, Organisation for Economic Co-operation and Development and International Energy Agency, Paris. 2008.

[4] IEA, *Contribution of Renewables to Energy Security*, Organisation for Economic Co-operation and Development and International Energy Agency, Paris. 2007.

[5] Awerbuch, S. and Berger, M., *Energy Security and Diversity in the EU: A Mean- Variance Portfolio Approach*, IEA Report Number EET/2003/03.

[6] Awerbuch, S., *The Role of Wind Generations in Enhancing Scotland's Energy Diversity and Security: A Mean-Variance Portfolio Optimization of Scotland's Generating Mix; Volume II: Detailed Analysis and Conclusions*. Prepared for Airtricity, Greenock, Scotland, 2005.

[7] Bates White, LLC, *A Mean-Variance Portfolio Optimization of Achieving California's Generation Mix to 2020: California's 33 Percept Renewable Portfolio Standard Goal*. Prepared For: California Energy Commission. 2007.

[8] IEA/OECD, *Projected Costs of Generating Electricity*, 2005.

[9] ASTM, Standard Practice for Measuring Life-Cycle of Buildings and Building Systems, Annual Book of ASTM Standards, Vol. 04.11.

[10] USDOE, *Solar Energy Technologies Multi Tear Program Plan, 2007-2011*, 2007.

[11] USDOE, *Solar Energy Technologies Multi Tear Program Plan, 2008-2012*, 2008.

[12] Oliver, M. and Jackson, T., The market for solar photovoltaics. *Energy Policy* **27**, pp. 371–385, 1999.

[13] Oliver, M. and Jackson, T., The evolution of economic and environmental costs for crystalline silicon PVs. *Energy Policy* **28** (14): 1011–1021, 2000.

[14] Oliver, M. and Jackson, T., Energy and economic evaluation of building-integrated photovoltaics, *Energy* **26** (4): 431-439, 2001.

[15] Lesourd, J-B, Solar photovoltaic systems: the economics of a renewable energy resource, *Environmental Modelling & Software* **16** (2): 47-156, 2001.
[16] Maycock, P., Cost reduction in PV manufacturing. Impact on grid-connected and building-integrated markets. *Solar Energy Materials and Solar Cells* **47**: 37–45, 1997.
[17] Sinden, G., "Characteristics of the UK wind resource: Long-term patterns and relationship to electricity demand", *Energy Policy* **35**: 112-127, 2007.

New benefit-cost methodology for evaluating renewable and energy efficiency programs of the US Department of Energy

R. T. Ruegg[1] & G. B. Jordan[2]
[1]*TIA Consulting, Inc., USA*
[2]*Sandia National Laboratories, USA*

Abstract

This paper describes a new methodology developed in 2009 for performing retrospective benefit-cost studies of Energy Efficiency and Renewable Energy (EERE) research and technology development (R&D) programs of the US. Department of Energy (DOE). The methodology uses a four-part benefits framework that includes economic, environmental, security, and knowledge benefits, and a technology cluster approach to address larger parts of major programs or entire programs. It improves on and extends an earlier approach developed by the US National Research Council (NRC) and applied in a 2001 NRC study, *Energy Research at DOE: Was It Worth It?*

The new EERE methodology was designed to answer the following questions about the EERE programs: To what extent have the programs thus far produced economic benefits in terms of resource savings relative to program costs? To what extent have the programs yielded environmental benefits, with a focus on health benefits from reduced air emissions? To what extent have the programs yielded energy security benefits in terms of reducing imported oil and reducing threats to the US energy infrastructure? To what extent have the programs built a knowledge base within each respective field and outside those fields? What has been the return on public investments in these energy programs thus far?

The new EERE methodology set forth in a draft Guide was applied in 2009 in four benefit-cost cluster studies to address the key evaluation questions in the following EERE program areas: Wind Energy, Solar Energy, Geothermal Energy, and Vehicle Technologies. This paper describes the methodology and gives an overview of its initial applications.

Keywords: evaluation, benefit-cost cluster study, energy, environment, security, knowledge.

WIT Transactions on Ecology and the Environment, Vol 131, © 2010 WIT Press
www.witpress.com, ISSN 1743-3541 (on-line)
doi:10.2495/EEIA100091

1 Introduction

Since 1978, DOE's research and development (R&D) programs in energy efficiency and renewable energy (EERE) have achieved many technical successes that have resulted in commercialized technologies and products found in today's markets, but most of these programs had not yet had independent assessments of returns on their R&D investments as of 2009. The last economic assessment of EERE programs was a 2001 NRC study [1]. At the same time, it was recognized by program managers that a major Federal energy program which has demonstrated benefits determined through systematic retrospective evaluation is better positioned to communicate its value to Congress, stakeholders and the public, than one which has not.

In 2009, EERE program staff set about to improve and extend the NRC evaluation approach based on recommendations made by reviews of the NRC study, and to apply it to a selection of EERE programs and subprograms. Goals were to develop a consistent, modified NRC approach for determining realized economic and other net benefits that would achieve the following: (1) model government additionality in detail, on a case-by-case basis, (2) refine and expand environmental benefits, particularly health benefits from reduced air pollution, (3) estimate security benefits as feasible, (4) expand the quantitative treatment of knowledge benefits, and (5) calculate returns to a program cluster, i.e., a whole EERE program or subprogram, rather than to a single project.

To this end, a new approach was developed, a draft "how-to" Guide [2] was prepared to implement the approach, experienced evaluators were identified, and four initial benefit-cost cluster studies were commissioned for completion early in 2010. After a detailed review by experts, recommended modifications are to be made to the methodology, and a final version of the Guide issued. This paper captures development through preparation of the draft reports; the related presentation will update developments through the review and resulting modifications in the reports and the Guide.

2 The evaluation framework

The evaluation framework used for the new EERE benefit-cost studies allows for a more comprehensive treatment than traditionally provided by benefit-cost assessments of energy programs.

2.1 Four categories of net benefits

As illustrated by table 1, there are four categories of benefits and costs included, rather than a focus only on savings and cost from changes in use of energy and other resources. The first row, net economic benefits, is expected to be estimated primarily in monetary terms. Program costs are included as an offset to benefits. The second row, net environmental benefits, is expected to include a monetary estimate of health effects associated with any reductions in air emissions, as well as non-monetary, quantitative measures of changes in green house gases, and non-monetary quantitative and qualitative treatment of any other environmental effects. The third row, net security benefits, at this time is expressed as equivalent barrels of imported oil avoided. The fourth row, knowledge benefits, includes quantitative, non-monetary estimates, derived mainly from patent and publication bibliometric techniques.

Table 1: Net benefits matrix.

Category of benefit and cost	Realized (retrospective) net benefits
economic	monetary + qualitative
environmental	monetary + physical units of energy + number of deaths avoided and other health measures + qualitative
security	physical units of energy + qualitative
knowledge	bibliometric measures

The values of each of the four categories of net benefits are to be presented separately. Then, an estimate is to be provided that combines the monetary results of net economic benefits (first row) with monetary estimates of health benefits (second row). For the 2009 studies, this is the only combined presentation of monetary benefits that is to be provided, due to greater uncertainty in attempting to estimate the other categories in monetary terms. Future benefit-cost studies may be extended to include additional monetary estimates of other categories of net benefits.

2.2 Cluster approach

A cluster approach compares benefits of selected elements of a defined technology area to investment costs of the entire associated program or major sub-part of it. The purpose is to provide an estimate of the minimum return for the whole program or major sub-part, without performing detailed analysis of all of its funded research projects or technologies. The approach works well for high-risk R&D programs where a few projects tend to be the big winners and investment in an array of projects is necessary to find successful ones. It is a potentially cost-effective approach to demonstrate that benefits from only a few elements in a cluster more than offset total program cluster investment costs.

The retrospective cluster approach begins with identifying a program or subprogram (i.e., a cluster) of evaluative interest. Next, a few technologies/ projects within the cluster are identified that appear to be among the more successful technically and commercially, and these are selected for detailed benefit-cost analyses. Those not selected for detailed treatment are treated qualitatively, including negative effects, if any. Finally, combined benefits of the technologies evaluated in detail are compared against entire cluster cost, and the results are conditioned by the qualitative results. Fig. 1 illustrates the comparison of benefits for selected elements of a technology cluster to costs of the elements and to entire cluster costs.

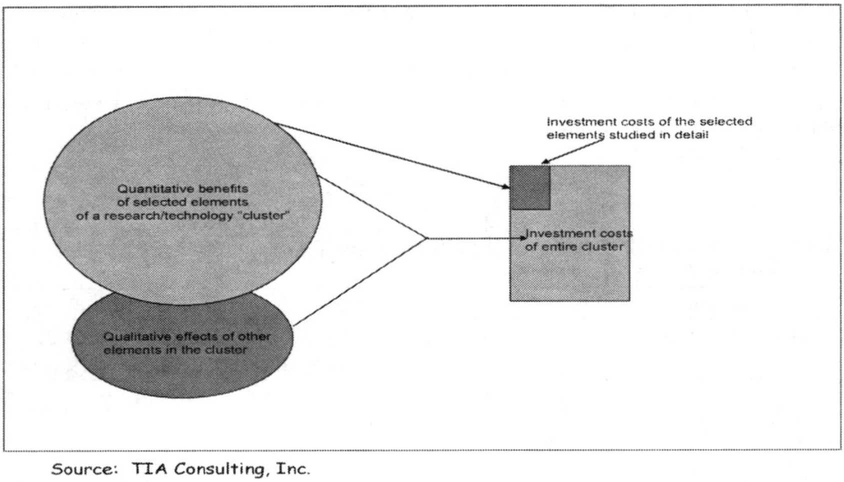

Figure 1: Comparing benefits of a technology cluster to entire cluster costs.

2.3 Estimating economic benefits and costs

2.3.1 Mansfield model provides a theoretical anchor for estimating returns

A model developed by the late economist Prof. Edwin Mansfield [3] serves as a unifying model across studies for valuing private and social economic returns from investment in new technology. Mansfield's approach includes market spillover effects which occur as others in the same industry as the innovator, within competitive markets, use the innovator's knowledge to imitate the innovation and drive down prices to consumers. Included are effects on customers of the investing/innovating firm and final consumers of related products and services within the industry. Not included are effects that occur as firms outside the innovator's industry draw from the same knowledge base to produce other goods and services in other industries. Also not included are more general effects of an enhanced knowledge base on the capacity to innovate in other areas.

2.3.2 Comparing a new technology in the cluster against the next best alternative

The merits of a new technology are judged against the next best alternative, i.e., the best choice that could be made in lieu of choosing the new technology. For a retrospective benefit-cost analysis, the next best alternative is defined by looking back to the time the investment decision was made for the new technology. There are several factors that affect the selection of the next best alternative that may help to inform the selection across studies in a consistent manner. One of these factors is whether the investment decision was constrained or unconstrained, that is, whether the choice was restricted, such as by regulatory requirements, or completely open. Another factor is whether the technology is new to the world or an improvement over an existing system. Yet another factor is whether the technology is a total system or a component of a system or a process used to make a system or component. The Guide provides an aid for defining the next best alternative for comparison.

2.3.3 Determining program additionality for each technology

A keenly important aspect of estimating the return on EERE's investment, i.e., the "return on public investment," is to provide evidence-based analysis of additionality. This entails delineating the part of benefits from the cluster technologies that is attributable to the cluster costs, and documenting evidence of cause and effect. The public program in question, for instance, may have accelerated technology entry into the marketplace; it may have improved the performance characteristics of the technology; it may have changed the technology's cost; it may have increased market size; it may have had other effects. The Guide provides an aid for organizing the additionality analysis and for mapping attribution to a technology timeline to show when and how an identified effect is estimated to have occurred.

Potential rival explanations of the estimated benefits must be addressed, such that it is the Program's effect that is identified in the additionality assessment and not other causes. Eliminating rival explanations is important because otherwise the benefits claimed for the Program could be due to other factors. For example tax credits may constitute a rival explanation for market expansion of a renewable energy technology – in opposition to an explanation that the market expansion resulted from R&D-induced advances in system performance.

2.3.4 Computing measures of economic performance

A positive public return means that part of societal benefits is attributable to EERE's program and that those attributed benefits exceed EERE's program cluster cost. The selected economic performance measures shown in fig. 2 – NB, B/C, and IRR – are used to provide estimates of the economic impact of the EERE program clusters. Results are computed for two discount rates – 3% and 7%, both defined as real rates, exclusive of inflation, in accordance with Federal guidance [4, 5]. Sensitivity analysis is performed by testing the outcome to alternative plausible values of other key variables.

The economic performance measures are computed based on monetary estimates about which the confidence level is relatively high. For the 2009 studies, these include the monetary estimates of economic impact and the monetary estimates of health effects from the environmental effects of reduced emission of certain air pollutants.

2.4 Estimating environmental benefits and costs

The focus of the quantitative estimation of environmental benefits in the EERE benefit-cost studies is on (1) estimating Green House Gas (GHG) effects, and (2) estimating public health benefits (i.e., avoided mortality, morbidity, and related costs) of reducing air emissions from fossil-fuel combustion. Effects on water

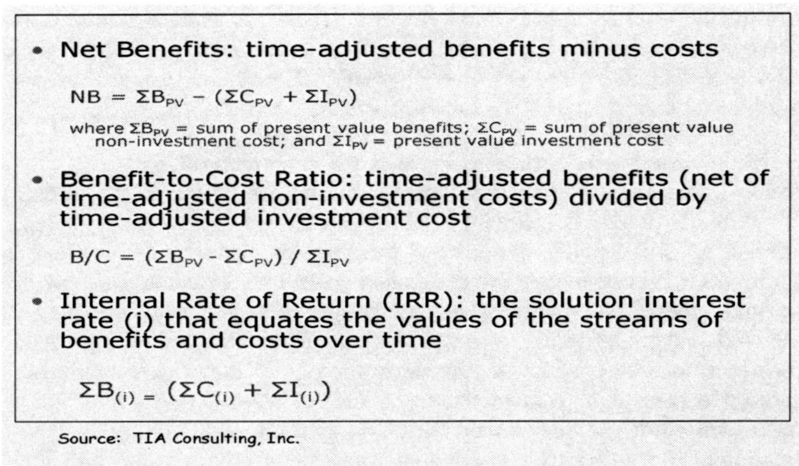

Figure 2: Three measures of economic performance.

resource use, water discharges, land resource use, and solid waste generation, if significant, are treated qualitatively.

2.4.1 Estimating Greenhouse Gas (GHG) emissions

Avoided GHG emissions from reduced combustion of fossil fuels, an important goal for EERE, is an aspect of environmental effects to be covered by the 2009 studies, with attention to carbon dioxide (CO_2), methane (CH_4), and nitrous oxide (N_2O). The 2009 draft study Guide recommends the use of the EPA Greenhouse Gas Equivalencies Calculator (available at www.epa.gov /cleanenergy/energy-resources/calculator.html) [7] to assist in assessing the consequences of the GHG effects.

2.4.2 Estimating health effects

To estimate health effects from changes in air pollution emissions attributed to the program cluster evaluated, the US Environmental Protection Agency's (EPA) COBRA model (Co-Benefits Risk Assessment Model, described in US EPA [6]) is used. To apply COBRA, it is necessary to enter the estimated changes in air emissions of particulate matter (PM), sulphur dioxide (SO_2), nitrogen oxide (NO_x), and volatile organic compounds (VOCs) into the model. Because not all air pollutants are taken into account by the model, the results obtained from using COBRA for the analysis is taken as a lower bound estimate of impact of health effects and their economic value. Table 2 shows the health effects included in COBRA, by type of effect. The model provides estimates of the incidence of each type of effect and related healthcare costs.

2.4.3 Treating other environmental effects

For other environmental effects – such as changes in water consumption effects, water discharge, land resource use, and solid waste generation – the 2009 draft

Table 2: Health effects included in COBRA.

Health effect	Description
Mortality	Number of deaths
Chronic bronchitis	Cases of chronic bronchitis
Non-fatal heart attacks	Number of non-fatal heart attacks
Respiratory hospital admissions	Number of cardiopulmonary-, asthma-, or pneumonia-related hospitalizations
Cardio-vascular related hospital admissions	Number of cardiovascular-related hospitalizations
Acute bronchitis	Cases of acute bronchitis
Upper respiratory symptoms	Episodes of upper respiratory symptoms (runny or stuffy nose; wet cough; and burning, aching, or red eyes)
Lower respiratory symptoms	Episodes of lower respiratory symptoms: cough, chest pain, phlegm, or wheeze
Asthma emergency room visits	Number of asthma-related emergency room visits
Minor restricted activity days (MRAD)	Number of minor restricted activity days (days on which activity is reduced but not severely restricted; missing work or being confined to bed is too severe to be MRAD)
Work days lost	Number of work days lost due to illness
Asthma exacerbations	Shortness of breath, wheezing, and coughing (in asthmatic individuals)

study Guide recommends the provision of data on physical units together with commentary description and explanation if feasible, and a qualitative treatment if quantitative estimates are not feasible.

2.5 Estimating security benefits and costs

Security benefits are attributed to reducing disruptions in the nation's energy supply. They also are attributed to reducing threats to the nation's energy infrastructure. In addition, and in the longer run, national security benefits may also result from reducing GHG emissions, by avoiding the host of overwhelmingly negative long-range national security consequences that have been predicted in response to global warming. These effects also are extremely difficult to assign values – particularly economic values – with any confidence. Associations among changes in energy efficiency, energy supply, energy prices, and security impacts involve many assumptions, with causal relationships far more uncertain than for those entailed in estimating the other categories of benefits included in the 2009 studies and addressed by the related draft Guide. Attempts at monetary valuation of those benefits would be subject to far greater margins of error than for the other monetary estimates contained in the studies. For this reason, the 2009 recommended EERE approach is to avoid monetary estimates of security benefits, and, to the extent feasible, to use an estimate of the reduction in physical units of barrels of oil equivalent deriving from use of renewable energy, increased efficiency, and energy conservation as a rough indicator of security benefits. In addition, if the technology cluster has knowable implications for the security of the nation's energy infrastructure, a qualitative description of these effects is to be provided.

2.6 Estimating knowledge benefits

The creation and dissemination of knowledge outputs are central to EERE's programs. These knowledge outputs embody the results of program R&D in papers, patents, presentations, models, resource maps, prototypes, technology demonstrations, test data, research tools, trained and experienced people, and networks of researchers working collaboratively. The take-up and use of these knowledge outputs by industry enables the production of more energy efficient and environmentally friendly products and new and improved renewable energy systems. Moreover, the acquisition of EERE knowledge outputs by the broader community increases interest in and willingness to adopt energy innovations, and enables researchers in other organizations to make further advances.

The knowledge base created by an EERE program or subprogram is more extensive than that captured by the technology-specific cases of the corresponding benefit-cost analysis. Therefore, each study incorporates an assessment of the program/subprogram's knowledge creation and dissemination. Techniques used to document knowledge creation and flow include bibliometrics (patent citation analysis and publication co-author and citation analysis); analysis of documents and databases; and interviews with experts.

Patent analysis has been used extensively to trace technological developments and is emphasized in the assessment of knowledge benefits. The analysis is based on the idea that the prior art embodied in a patent referenced by a later patent provides part of the foundation for the later invention. A correlation between patent citations and measures of technological and scientific importance has been documented; highly cited patents tend to contain technological information of particular interest or importance. A summary of validation studies supporting patent analysis for assessing knowledge benefits and dissemination is found in Breitzman and Mogee [8]. Backward patent tracing is used to determine the extent to which DOE-funded research in the program/subprogram area has formed a foundation for technologies in the target area developed by leading commercial innovators in the industry. Forward patent tracing is used to investigate the impact of DOE-attributed patents resulting from the program/subprogram on subsequent technological developments, regardless of where they occur – whether in or outside the technology and industry area of primary EERE program interest.

The knowledge sections of the four 2009 benefit-cost studies were derived from four separate studies by Ruegg and Thomas [9–12], which traced linkages from the outputs of EERE's R&D programs to downstream developments.

3 Essential study characteristics

The draft methodology Guide drew on the following sources in developing a list of essential study characteristics: a report from the American Evaluation Association (AEA) Task Force on Guiding Principles for Evaluators [13]; a White House guidance memo to heads of agencies and executive departments on emphasizing evaluation [14], other evaluation resources, and EERE-stated preferences for uniformity in report format. The following provides a non-exhaustive list of essential study characteristics:

- Clear statement of evaluation study's design and objectives; appropriate design given the objectives; and appropriate objectives given the nature and stage of the program
- Clear statement of benefit-cost framework and conceptual models used
- Clear description of the program cluster, its components, logic, and cost
- Clear account of the technologies selected for detailed case study, rationale for selection, and relationship of those selected to the larger cluster
- Appropriate designation of the next-best alternative to use as a baseline for estimating the differential effects of each technology selected
- Assessment of the context and external influences that may constitute rival explanations of outcome; adequate control for rival explanations of outcomes
- Use of valid protocols and procedures in data collection
- Adequate identification/documentation of data quality and related issues of uncertainty; inclusion of discussions of levels and sources of uncertainty in the study report; and appropriate reflection of uncertainty in the analyses
- Critical assumptions are stated and documented, and study limitations are identified
- Systematic and appropriate analyses to achieve objectives within the conceptual framework
- Findings are evidence-based. Conclusions are in sync with findings. Implications flow from findings and conclusions, consistent with methodology.
- Findings are conservative, in that they likely understate the actual benefits from the cluster investment
- Evaluation objectives are achieved.

4 The 2009 benefit-cost studies

DOE/EERE commissioned four benefit-cost studies in 2009, to be conducted according to the new draft EERE methodology. These initial benefit-cost studies [15–18] are listed in table 3, identified by the EERE program or subprogram evaluated, by authors and their organizational affiliation, and the group of technologies of focus:

Table 3: Four initial benefit-cost studies performed according to EERE's new draft methodology.

Program/subprogram evaluated	Authors	Author affiliation	Technologies assessed in detail
Advanced Combustion Engine Technologies (part of the Vehicle Technologies Program)	A. Link	University of North Carolina at Greensboro	-Laser and optical diagnostic technologies related to heavy-duty diesel engines
Geothermal Program	M. Gallaher A. Rogozhin J. Petrusa	RTI International	-Polycrystalline diamond compact (PDC) drill bits -Binary cycle power plant technology -TOUGH series of reservoir models -High-temperature geothermal well cements
Solar Photovoltaics (part of the Solar Energy Technologies Program)	A. O'Connor R. Loomis F. Braun	RTI International	-Flat-plate solar array -Photovoltaic manufacturing technology (PVMaT) -Thin film PV
Wind Energy Program	T. Pelsoci	Delta Research Co.	-Turbulence models -Aerodynamics and design codes -Variable speed drives -Blade materials characterization -Airfoil design codes -Demonstration and testing

These reports are in review as this paper is prepared. Following the detailed, expert reviews of the reports, they will be revised accordingly, and published. It is expected that summary results will be available for release as part of the Conference presentation.

5 Next steps

The Guide on the new EERE benefit-cost methodology was left in draft form pending completion and review of the initial set of benefit-cost studies and feedback from the researchers who conducted the studies. That feedback, together with the already completed extensive reviews of the draft Guide by an external review panel and internal DOE reviewers, as well as current EERE requirements, will inform the need for further modifications of the Guide prior to publication.

Acknowledgements

Work presented here was completed for the US Department of Energy (DOE) Office of Energy Efficiency and Renewable Energy by and under contract to Sandia National Laboratories, Albuquerque, New Mexico, USA under Contract DE-AC04-94AL8500. Sandia is operated by Sandia Corporation, a subsidiary of Lockheed Martin Corporation. The views and opinions of authors expressed herein do not necessarily state or reflect those of the United States government or any agency thereof.

References

[1] U.S. National Research Council (NRC) and applied in a 2001 NRC study, *Energy Research at DOE: Was It Worth It?*

[2] Ruegg, R. & Jordan, G. *Guidelines for conducting EERE retrospective benefit-cost studies*, (draft for internal DOE use only), September 20, 2009.

[3] Mansfield, E. *Estimating social and private returns from innovation based on the Advanced Technology Program*, 1996.

[4] OMB (Office of Management and Budget). *Circular no. A–4: Regulatory analysis, 2003.*

[5] OMB (Office of Management and Budget). *Circular no. A–94: Guidelines and discount rates for benefit–cost analysis of federal programs*, 1992.

[6] U.S. EPA (Environmental Protection Agency). *User's manual for the Co-Benefits Risk Assessment (COBRA) screening model*. Developed by Abt Associates Inc., June 2006.

[7] U.S. EPA. EPA Greenhouse Gas Equivalencies Calculator (available at www.epa.gov/cleanenergy/energy-resources/calculator.html).

[8] Breitzman, A. & Mogee, M., The many applications of patent analysis, *Journal of Information Science*, 28(3), pp. 187-205, 2002.

[9] Ruegg, R. & Thomas, P. *Linkages from EERE's wind energy program R&D to commercial renewable power generation*, U.S. Department of Energy, September 2009, www1.eere.energy.gov/ba/pba/pdfs/wind_energy_r_and_d_linkages.pdf.

[10] Ruegg, R. & Thomas, P. *Linkages from EERE's geothermal R&D to commercial power generation*, U.S. Department of Energy, March 2010, in press.

[11] Ruegg, R. & Thomas P. *Linkages from EERE's advanced combustion R&D to improved vehicle energy performance*, U.S. Department of Energy, March 2010, in press.
[12] R. Ruegg & P. Thomas, *Linkages from EERE's solar photovoltaic R&D to commercial renewable power from solar energy*, U.S. Department of Energy, March 2010, in press.
[13] American Evaluation Association (AEA), *Guiding Principles for Evaluators*, www.eval.org/Publications/GuidingPrinciples.asp.
[14] Executive Office of the President, Office of Management & Budget, memo for heads of executive departments and agencies, *Increased emphasis on program evaluation*, October 7, 2009.
[15] Pelsoci, T. M. Retrospective benefit cost study of U.S. DOE Wind Energy Program Technology Investments, U.S. Department of Energy, March 2010, in press.
[16] Gallaher, M., Rogozhin, A., & Petrusa, J. *Geothermal technology benefit cost analysis*, U.S. Department of Energy, March 2010, in press.
[17] Link, A. An economic evaluation of DOE solar program investment in photovoltaic energy systems, U.S. Department of Energy, March 2010, in press.
[18] O'Connor, A. C., Loomis, R. J. & Braun, F. M. An economic evaluation of DOE solar program investment in photovoltaic energy systems, U.S. Department of Energy, March 2010, in press.

Assessing the efficiency of municipal expenditures regarding environmental protection

J. Soukopova & E. Bakos
Department of Public Economics,
Faculty of Economics and Administration, Masaryk University,
Czech Republic

Abstract

This article deals with the efficiency of the usual municipal expenditures on environmental protection and suggests a methodology for assessing this efficiency. Firstly, the paper analyses the concept of efficiency from the view of individual rationality. The authors consider efficiency in the sense of 3E methodology – economy, efficiency and effectiveness and the methodology of sustainable development – the social, environmental and economic parts of sustainable development, as well as the role of those who make decisions in environmental politics. A proposal of a methodological procedure for assessing municipal expenditure efficiency is presented next. It uses multi-criteria assessment, where a dominant criterion of performance is C/E. This procedure is applied to a file of environmental expenditure data from the representative sample of municipalities in selected areas of environmental protection that were used in a project of the Ministry of Environment of the Czech Republic SP/4i1/54/08 "Analysis of municipal budgets efficiency in relation to the environmental protection". The data comes from selected municipality budgets and are analyzed for the time range of 2001-2008, because the data has been in electronic form since then.

Keywords: efficiency, effectiveness, economy, municipal environmental expenditures, sustainable development.

WIT Transactions on Ecology and the Environment, Vol 131, © 2010 WIT Press
www.witpress.com, ISSN 1743-3541 (on-line)
doi:10.2495/EEIA100101

1 Introduction

The issue of the relation between economic growth and environmental protection has become increasingly important in recent years. In question are also the effects of environmental policy in individual regions and the influence of environmental policy on economic growth and other basic regional economic indicators, such as unemployment, inflation, trade and living standards. The problem of allocation of public expenditures in this field is also related with this; specifically how much, in what way and for what purpose should taxpayers' money be spent in relation to environmental protection. This was the reason for the Ministry of the Environment (MoE) of the Czech Republic funding project SP/4i1/54/08 "Analysis of local budgets and their efficiency in relation to environmental protection". Its main objective is to evaluate the efficiency of public expenditures and other financial instruments in the field of environmental protection with focus also on particular regions and the optimization of the incidence of public subsidies in the field of environmental protection on macro- and micro-economical levels. The important part is identification of factors influencing the absorption capacity of individual regions in the Czech Republic and the setting of indicators for the evaluation of their efficiency.

2 Analysis of environmental public expenditures

Public expenditures in the field of environmental protection are the important part of total public expenditures and probably even in times of financial crisis their amount will not decrease notably, thanks to the active policy of the European Union and expenditures from its structural funds. Figure 1 shows the progression of public expenditures since 1997.

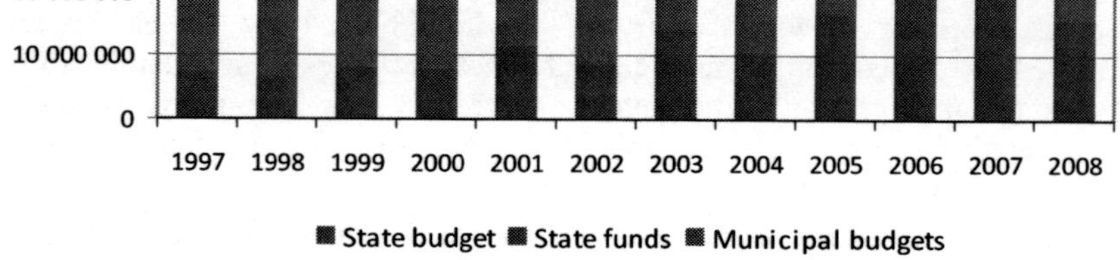

Figure 1: Environmental expenditures of public budgets (in thousands CZK) [11].

In Figure 1 we can see that municipal expenditures made throughout the time covered are always more than 50% of total environmental public expenditures.

Environmental expenditures in the budget structure are divided according to the Classification of Environmental Protection Activities and Expenditure (CEPA 2000), which differentiate the protection of ambient air and climate, wastewater management, waste management, protection and remediation of soil, groundwater and surface water, noise and vibration abatement, protection of biodiversity and landscapes, protection against radiation, research and development and other environmental protection activities [12]. As shown in Figure 2, the largest parts of environmental municipal expenditures are wastewater management expenditures, waste management expenditures and protection of biodiversity landscapes expenditures.

3 Environmental public expenditures efficiency

One of the biggest problems of contemporary economic theory is the one of defining and measuring the efficiency, or in other words use, of resources and their transformation into outputs and outcomes. Already in 1957 Farell had asked the question of how to measure efficiency and pointed out [8] its importance for economic policy makers: "*it is important to know how far a given industry can be expected to increase its output by simply increasing its efficiency, without absorbing further resources*" [4]. There have been several decades of efficiency evaluation and technologies are greatly improved and advanced. However, there still remains a conceptual challenge in relation to public expenditures, given that the problem is also complicated by the fact that outcomes of the public sector are

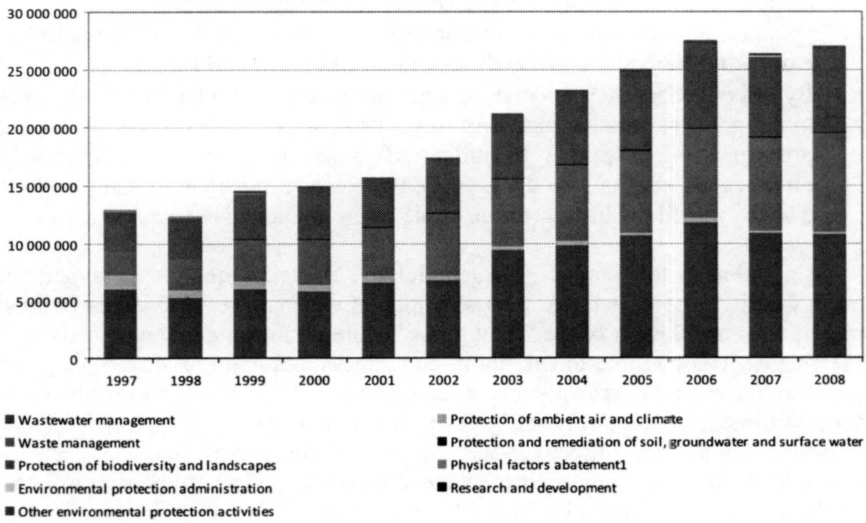

Figure 2: Municipal environmental expenditures according to CEPA 2000 (in thousands CZK) [11].

often off-market, lacking relevant data and thus making them unquantifiable, as stated by the collective of authors at the European Commission [8]. It is the very conceptual frame of inputs, outputs and outcomes that these authors are pointing out. They highlight the difference in comprehension concepts of output and outcome. While they see the efficiency in the transformation of inputs to outputs (comparing it to productivity, which they see as a level of product created from input used), which includes also the concept of the production possibilities frontier (in other words the more output we create from a given input or the less input is required for desired output, the more efficient is the activity), they ask for effectiveness in relation between output and outcome, which they perceive as richness or growth in society and which is, besides political decisions, influenced by various other external factors (those identified by member states as key factors related to public expenditures were performance orientation, organizational aspects, human resource management, information technology utilization). The above described problem of expressing differences between concept of inputs, outputs and outcomes and related understanding and measuring of efficiency also related to public expenditures has been investigated by Mandl et al. [8] and many other authors [3, 5–7, 9].

To evaluate the efficiency of public expenditures (environmental), most authors use the methodology of 3E – economy, efficiency and effectiveness, which they perceive from a theoretical basis as follows:

- By economy they understand such use of public expenditures that leads to provision of the given objectives with the least amount of resources spent, while keeping up to the corresponding quality of tasks;
- By efficiency they understand such use of public expenditures that acquires the greatest possible amount, quality and contribution to the given objectives compared to the amount of resources spent in order to fulfil them.

Economy and efficiency are for purposes of quantification and in respect of usage of methods of economic analysis understood as cost efficiency.

- By effectiveness they understand such use of public expenditures that leads to the greatest possible output in respect of the desired outcome, which is a prerequisite for optimal fulfilment of goals set in advance. Therefore, effectiveness means how the produced goods or services (for example waste disposal) fulfil the utility (for example clean municipal environment without waste).

In addition to this classic 3E methodology, the term ‘quality’ is sometimes used. Quality means such that the use of public expenditures provide an optimal rate of accomplishment of the “right goals” while performing given objectives. It means that it is possible to ask about correctness and appropriateness of given goals in, for example, strategic documents or from the point of the legitimacy of their fulfilment, or the utility set by them. It is important to strongly differentiate between quality and effectiveness, which are sometimes interchanged, for example in the concept of quality, where it comes to optimal fulfilment of goals while carrying out given objectives. In this concept it is not clear enough what process is used to set up goals and to what extent these goals are “objectively” right, or appropriate. Sometimes it is possible to purposefully (in terms of purpose) fulfil the goals, but not in optimal ways, meaning not taking into account cost amount. When judging all these criteria (economy, efficiency, effectiveness and quality) we can speak of the economical efficiency of public expenditures. For the complex view we need to add that sometimes we distinguish between the terms technical and allocation efficiency. However, this concept's analysis is beyond the scope of this text. Figure 3 shows the concept of economical efficiency, from which we move out into further analysis and we use it for the construction of a methodology for the evaluation of environmental municipal expenditures.

4 Methodology of efficiency evaluation

One of the contemporary problems is how to allocate public expenditures in the field of environment protection more effectively.

When considering efficiency, the methodology is based on a multi-criteria evaluation of efficiency based on three basic pillars of sustainable development. When we designed the methodology we started from the assessment of efficiency in terms of social, environmental and economical points of view (see Figure 4).

4.1 Social aspect of evaluation

The social criteria of evaluation take the social aspect of existing expenditure into account. The complex criterion for evaluating efficiency from the social point of view K_S could be constructed as follows:

$$K_S = \sum_{i=1}^{n} w_i k_{Si}, \tag{1}$$

where k_{Si} is the social efficiency criterion (in percent),
n is the number of criteria,
w_i is the weight of criterion No. i, and $\sum_{i=1}^{n} w_i = 1$.

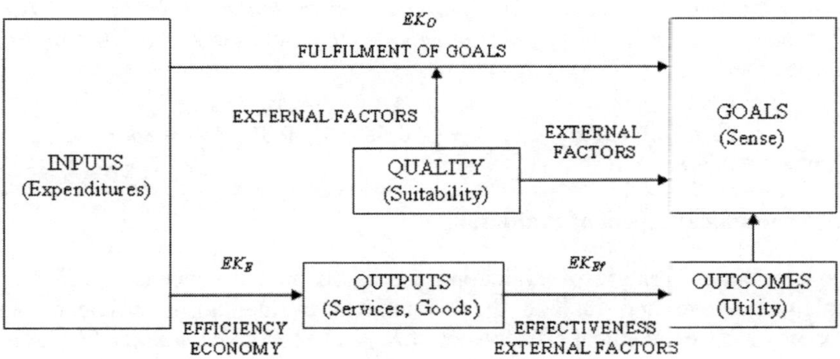

Figure 3: Conceptual conception of efficiency of public expenditures.

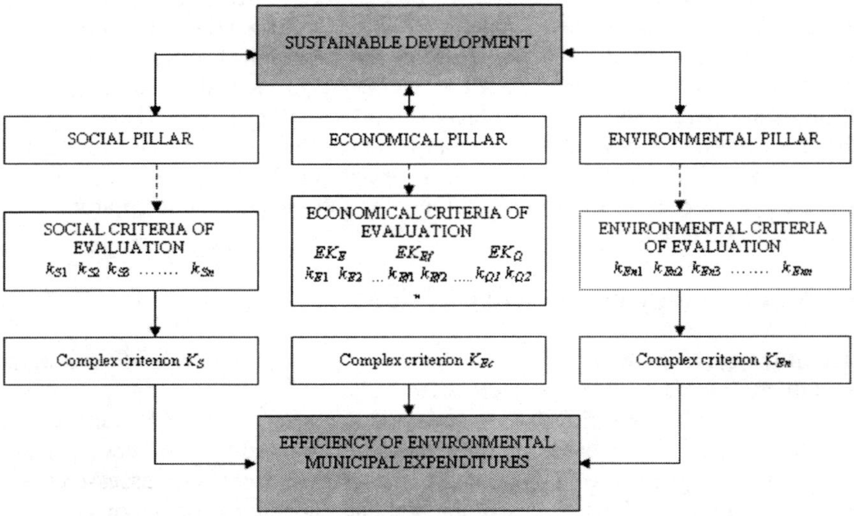

Figure 4: Scheme of environmental public expenditure efficiency evaluation.

If it holds that $0 \leq K_S \leq 1$ and if K_S = 0 then the expenditure is absolutely inefficient.

Example 1

When it comes to municipal waste management expenditures, suitable criteria for social efficiency evaluation of given expenditures could be the following:

k_{S1} *Willingness to sort municipal waste (in percent)*

k_{S2} *Employment – Influence on employment (is the given service carried out by a local company or an external one, and so on) (in percent)*

k_{S3} *Living standard of citizens – has the expenditure had a positive impact on the living standard of citizens in the municipality (in percent)*

When evaluating municipal waste management expenditures in Brno, experts gave these weights to the given criteria $w_1 = 0.4$ $w_2 = 0.3$ $w_3 = 0.3$ *and the following values:*

Criterion	k_{S1}	k_{S2}	k_{S3}
Criterion value	0.58	0.85	0.86

Then, $K_S = 0.748$.

4.2 Economical aspect of evaluation

The economical criteria of evaluation are based on the concept of efficiency explained above and include the economical evaluation of efficiency and economy EK_E, economical effectiveness EK_{Ef} and economical quality EK_Q, so:

$$K_E = EK_E + EK_{Ef} + EK_Q, \tag{2}$$

where K_E is the complex criterion of economical efficiency evaluation,
EK_E is the complex criterion of economical efficiency and economy evaluation (cost efficiency evaluation),
EK_{Ef} is the complex criterion of economical effectiveness evaluation,
EK_Q is the complex criterion of economical quality evaluation (quality of environmental goals).

A more detailed explanation of the methodology of evaluation according to the given complex criteria follows.

4.2.1 Economy and efficiency evaluation – EK_E

The most commonly used methods for evaluating the efficiency of public expenditures (capital and current) are Cost-minimization Analysis, Cost-effectiveness Analysis (CEA), Cost-utility Analysis (CUA) and Cost-benefit Analysis (CBA). These methods are suitable for the evaluation of the efficiency of public expenditures for environmental protection. The only exception is Cost-minimization Analysis, which only compares amount of costs (expenditures) in certain investment; therefore we will not consider it further for the evaluation of environmental public budget expenditures. The efficiency evaluation of current expenditures of public budgets, however, encounters several limitations. This is because current expenditures usually consist of expenditures for public services – services of common interest. This makes it quite difficult to evaluate expenses using CBA or CUA. In the case of CBA it is difficult to estimate the benefit of these services in terms of money and as for CUA, the situation is even more complicated because there is no suitable methodology for environmental expenditures (however it exists for healthcare and others) [1]. Therefore, the most exact method appears to be CEA [2]. When it comes to the evaluation of efficiency, and for the evaluation of C/E, choosing efficiency indicator E as a complex criterion created with the help of a multi-criterial analysis depending on the factors influencing expenditures on a given environmental service appears to be the best option.

Let K_E be a set of criteria for the evaluation of the quality of environmental public budget expenditures, where $K_E = (k_{E1}, k_{E1}, \ldots, k_{En})$, so

$$E = f(k_{E1}, k_{E2}, \ldots, k_{En}), \tag{3}$$

where k_{Ei} is the criterion of cost efficiency and economy evaluation,
n is the number of outputs for a given environmental expenditure.

Then the cost efficiency of a given expenditure could be expressed as follows:

$$CEA = \frac{C}{E} \geq 0 \rightarrow \min \tag{4}$$

where C are environmental expenditures,
E is the indicator of the cost efficiency evaluation.

If $CEA \leq 1$, then the expenditure is efficient, if $CEA > 1$, then the expenditure is inefficient. Because the criterion is minimizing, it needs to be transformed into a maximizing one. Therefore for the construction of the EK_E criterion we will use the following formula:

$$EK_E = \frac{1}{CEA} = \frac{E}{C} \geq 0, \tag{5}$$

where if $EK_E>1$, then the expenditure is efficient and $EK_E \to \max$.

Example 2
In 2008 the city of Brno spent in the area of waste management 189,947.87 thousand CZK for municipal waste collection and 176,511.6 thousand CZK for the use and disposal of municipal waste, i.e. the total cost of waste management is C = 366,459.47 thousand CZK. The same year the city of Brno was producing Q = 118,662.87 tons of municipal waste (k_{E1}), the average price for waste treatment was p = 1,500 CZK (incinerator) (k_{E2}), the average distance to processing facilities was v = 20 kilometres (k_{E3}), and the average size of a mean of transportation for waste was 25 t (k_{E4}), the average rate for transport was 45 CZK/t (k_{E5}) and the average rate for handling was 30 CZK/t (k_{E6}). Then, in the case of the collection of municipal waste being the criterion, the waste amount, price of waste manipulation, price of waste transport, means of transportation capacity and distance to processing facilities are used to calculate the costs of collection, as follows:

$$E = E_1 + E_2,$$

$$E_1 = N_S = 2 * v * s_d * \frac{Q}{k_d} * m, \quad E_2 = Q * p,$$

where v *is the distance from the facility (landfill, incinerator) [km] – (k_{E3})*
s_d *is the rate for transportation[CZK/km], considered 45 CZK/km (k_{E5})*
Q *is the amount of waste [t] (k_{E1})*
k_d *is the capacity of the means of transport for waste [t], considered to be a maximal capacity of 25 t. (k_{E4})*
m *is the price for manipulation [CZK/t]*

Then, E_1 = 192,233.85 CZK, E_2 = 177,994.305 CZK and E= 370,228.155. It follows: EK_E=1.01.

4.2.2 Evaluation of the effectiveness – EK_{Ef}

Let K_{Ef} be a set of criteria for the evaluation of effectiveness of environmental municipal expenditures, where $K_{Ef}= (k_{Ef1}, k_{Ef1}, \ldots, k_{Efn})$, then

$$EK_{Ef} = \sum_{i=1}^{n} w_i k_{Efi}, \tag{6}$$

Where k_{Efi} is the criterion determining the results of a given expenditure– percentual fulfilment of goal No. *i* (criterion acquires values 0-1),
n is the amount of outcomes (goals) for given environmental expenditure,
w_i is the weight of the *i*-numbered criterion, which fulfils $\sum_{i=1}^{n} w_i = 1$.

It holds that $0 \leq EK_{Ef} \leq 1 \to \max$.

Example 3
The City of Brno is planning in its waste management area the following objectives and performance criteria of expenditure effectiveness.

1. *Increase material utilization of municipal waste to 50% by 2010 compared to 2000* – $k_{Enf\,1}$;
2. *Material utilization of municipal waste in relation to the whole Czech Republic (ensure the collection, subsequent use or disposal of controlled hazardous components of municipal waste (50% in 2005 and 75% in 2010))* – $k_{Ef\,2}$;
3. *Ensure recycling of construction and demolition waste (utilize 50% of the weight of emerging construction and demolition waste before 31. 12. 2005 and 75% before 2012)* – $k_{Ef\,3}$;
4. *Prefer incineration of mixed municipal waste with energy recovery prior to landfill storage* – $k_{Ef\,4}$;
5. *Reduce the weight ratio of landfilled waste with the perspective of a further reduction of 20% in 2010 compared with 2000* – $k_{Ef\,5}$;
6. *Decrease the ratio of landfilled waste with energetic utilization potential (35% in 2010)* – $k_{Ef\,6}$;
7. *Decrease the ratio of landfilled biodegradable municipal waste (75% of what the production was from 1995 to 2010)* – $k_{Ef\,7}$;
8. *Increase the utilization of waste through recycling up to 55% in 2012* – $k_{Ef\,8}$.

For simplicity, all the criteria were assigned the same weight, $w_i = 0.125$. *The expert panel gave each criterion the following values:*

Criterion	k_{Ef1}	k_{Ef2}	k_{Ef3}	k_{Ef4}	k_{Ef5}	k_{Ef6}	k_{Ef7}	k_{Ef8}
Criterion value	0.95	1	0.86	1	0.85	0.95	0.65	1

Therefore, $EK_{Ef} = 0.9075$.

4.2.3 Evaluation of the quality – EK_Q.

Let EK_Q be a set of criteria for the evaluation of the quality of environmental public budget expenditures, where $EK_Q = (k_{Q1}, k_{Q1}, \ldots, k_{Qn})$, then

$$EK_Q = \sum_{i=1}^{n} w_i k_{Qi} \tag{7}$$

Where k_{Qi} is the criterion determining quality – quality of a given goal – in connection with strategic documents of region or state (in percent) (criterion acquires values 0-1),

n is the amount of outcomes (goals) for given environmental expenditure,

w_i is the standardized weight of criterion No. i.

Example 4
The South Moravian Region has in its strategic document – Waste Management Plan (WMP) – 25 goals related to waste management. The city of Brno has given in its waste management eight goals, which are all included in the WMP South Moravian region; therefore, these criteria take the value of 1 (100% associated with the strategic documents). Considering the evaluation of quality of

expenditures, it is possible to use the criteria in Example 3 and build EK_Q, when $EK_Q = 1$.

For the city of Brno the complex criterion for the evaluation of economical efficiency comes out as follows:

$$K_E = EK_E + EK_{Ef} + EK_Q = 0.9898 + 0.9075 + 1 = 2.8973$$

4.3 Environmental aspect of evaluation

Environmental criteria of evaluation are obtained from indicators of sustainable development in the selected field of environmental protection. The complex criterion of the evaluation of efficiency could be from the view of environmental K_{En}, constructed as follows:

$$K_{En} = \sum_{i=1}^{n} w_i k_{Eni} \tag{8}$$

Where k_{Eni} is the criterion of environmental efficiency, $k_{Eni} \rightarrow \max$
n is the amount of criteria,
w_i is the standardized weight of criterion No. i.

It holds that if $K_{En} \geq 0$. If $K_{En} = 0$, the expenditure is fully inefficient.

Example 5
Considering waste management expenditures, the criteria for the evaluation of environmental efficiency could be the following, which are maximizing:

k_{En1} *Amount of municipal solid waste per capita in comparison with the Czech national average (national average proportion of the municipality value);*

k_{En2} *Weight ratio of going to landfills, compared with the Czech average (ratio of Czech average to the actual municipality value);*

k_{En3} *Waste management expenditures per capita compared to the Czech average (ratio of Czech average to the actual municipality value);*

k_{En4} *Ratio of biodegradable municipal solid waste going to landfills, compared with the Czech average (ratio of Czech average to the actual municipality value);*

k_{En5} *Utilization of waste through recycling compared with the Czech average (ratio of Czech average to the actual municipality value).*

Experts assigned these criteria by similar weight of $w_i = 0.2$. The expert panel attributed to each criterion the following values:

Criterion	k_{En1}	k_{En2}	k_{En3}	k_{En4}	k_{En5}
Criterion value	1.099	1.541	0.823	1.125	1.02

Then, $K_{En} = 1.122$

4.4 Summary of the methodology

The sequence of our suggested methodology for the evaluation of public budget expenditures for environmental protection could be shown in several phases and steps:

1. Phase – evaluation of efficiency from the social view, $0 \leq K_S \leq 1 \rightarrow \max$;

2. Phase – the evaluation of the economical efficiency:
   - Step 1 – the evaluation of efficiency and economy of expenditures (whether the given goals are being fulfilled with minimal costs, or if the environmental benefits with given costs are maximized). $EK_E > 1 \rightarrow$ max;
   - Step 2 – the evaluation of effectiveness (how municipal environmental expenditure ensures the setting of goals). $0 \leq EK_{Ef} \leq 1 \rightarrow \max$;
   - Step 3 – the evaluation of quality (quality of goals is a crucial problem of expenditures, which is why we also evaluate it). $0 \leq EK_Q \leq 1 \rightarrow \max$;
3. Phase – the evaluation of efficiency from an environmental view. $K_{En} \geq 0 \rightarrow \max$.

Example 6
If we apply the methodology to the waste management expenditures of Brno in 2008, then we can use Examples 1–5 and the evaluation of efficiency according to our methodology would be as follows:

Phase 1		*Phase 2*		*Phase 3*	*Overall evaluation*
K_S	EK_E	EK_{Ef}	EK_Q	K_{En}	
0.905	1. 01	0.9075	1	1.122	4.9895

When it is compared with the average of municipalities of the South-Moravian region, where the overall evaluation value is 4.8254, we can say that the efficiency of Brno's waste management expenditures is very good.

5 Conclusion

This paper is one of the results of the project of the Ministry of Environment (MoE) of the Czech Republic SP/4i1/54/08 "Analysis of municipal budgets efficiency in relation to the environmental protection", where we identified that the efficiency evaluation of municipal environmental expenditures is an extraordinary difficult task. Just to determine economy and efficiency from a quantifying viewpoint with methods of economical analysis is not simple. The greatest problem is to estimate the benefits of public services in terms of money. We discussed why the most appropriate way seems to be the Cost-effectiveness Analysis and its application as a part of a multi-criteria analysis, depending on factors influencing expenditures on a given environmental service. Determination of all these factors, as shown in the examples in the paper, is a prerequisite for establishing an indicator of efficiency.

It is much more complicated when determining efficiency and quality of public expenditures. This opens several questions and tasks, which we are solving in the project No. SP/4i1/54/08 of MoE. What is the extent to which outputs are active in relation to the outcomes? How should one determine the success of the objectives? Are the goals set "correctly"? How should one identify that? How should one assess the quality of the given objectives? Are citizens' views and opinions relevant? Or it can be assumed using previously given objectives in the national and regional documents? For simplicity, we just assessed the compliance of the objectives set at the local level with the objectives set in national and regional strategic documents. We believe that this is one of the ways to assess the efficiency of public spending on environmental protection. The set of objectives and targets in the strategic documents of the Czech Republic and its regions, in our view, in itself reflects practical effects for improving the environment in the region and this leads to an increase in the overall living standard of the population and sustainable development.

At the same time we realize that the described problem in the project No. SP/4i1/54/08 of the MoE is much more complicated in practice, because the amount of public spending is influenced by a variety of external factors, such as orientation to performance, organizational aspects, human resources, the use of information technology, political decisions, interest groups, etc. Some of these factors cannot be quantified, they can only be described. International organizations already recognize the complexity of size and efficiency of public expenditures and their management to protect the environment and thus there have been formulated advices referred to as “good practices” [10] for the management of public expenditure with regard to environmental protection. These “good practices”, however, are more general and refer to a broader access to public spending than the presented methodology for the assessment of the public spending efficiency of local budgets for environmental protection in the paper.

Acknowledgement

This paper is supported by the project of the Ministry of Environment of the Czech Republic SP/4i1/54/08 “Analysis of municipal budgets efficiency in relation to the environmental protection”.

References

[1] Boardman, A., E. *Cost-benefit analysis: concepts and practice.* 2nd ed. Upper Saddle River: Prentice Hall, 2001. 526 p., ISBN 0-13-087178-8

[2] Levin, H., M., McEwan, P., J. *Cost-effectiveness analysis: Methods and applications.* 2nd ed. Sage Publications, Inc; 2000, 308 p., ISBN 0-7619-1934-1

[3] Ochrana, F. *Veřejný sektor a efektivní rozhodování.* 1. vyd. Praha: Management Press. 246 p. ISBN 80-7261-018-X, 2001

[4] Farrell, J. The Measurement of productive efficiency, *Journal of the Royal Statistical Society,* Part III Vol.120, pp.11ff, 1957

[5] Pavel, J., Slavíková L., Jílková J. Economics Instrument of Environmental policy: Costly Taxes and Low Effectiveness, *Journal of Economics*, vol. 57, No. 2, pp 132-144. ISSN 0013-3035, 2009

[6] Robinson, M. Output-Purchase Funding and Budgeting Systems in the Public Sector. *Public Budgeting & Systems*, vol. 22, pp. 17-33. ISSN 1096-3367, 2002
[7] Allen, R., Tommasi, D. (eds) Managing Public Expenditure: A reference book for transition countries, Paris, OECD, 2001
[8] Mandl, U., Dierx, A., Ilkowitz, F. The effectiveness and efficiency of public spending, European Commission, *Economic paper* 301, 2008
[9] Nath, N., Van Peursem, K., Lowe, A. Public Sector Performance Auditing: Emergence, Purpose and Meaning. *Working Paper Series – Department of Accounting*, The University of Waikato, 2005. no. 81, 40 p. ISSN 1173-7182
[10] Good practices of public environmental expenditure management in transition economies, Fifth Ministerial Conference, Environment for Europe, Kiev, Ukraine, 21-23 May 2003 www.oecd.org/dataoecd/51/59/34595093.pdf
[11] Information system ARIS http://wwwinfo.mfcr.cz/aris/
[12] EUROSTAT http://ec.europa.eu/eurostat/ramon/nomenclatures/index.cfm?TargetUrl=LST_NOM_DTL&StrNom=CEPA_2000&StrLanguageCode=EN&IntPcKey=&StrLayoutCode=HIERARCHIC&CFID=12332028&CFTOKEN=344c95f81e92cf39-AD596786-B0DD-5AA5-B6EB07E23EB60B0&jsessionid=f90093be44df7a11125a

Car scrappage incentives policies: a life cycle study on GHG emissions

M. Lelli, G. Pede, M. P. Valentini & P. Masoni
ENEA, Italy

Abstract

In the last decade a certain number of car scrappage incentives programs have been promulgated to promote earlier replacement of aging vehicles with new ones, cleaner and more efficient, with the aim of pollutant emission reduction and transport safety improvement. Some concerns have been raised about the effectiveness and efficiency of those measures, in particular in relation to the overall environmental impacts along the life of the vehicle phases – production, use, end of life – and fuel life chain.

This life cycle study is focused on the effects on greenhouse gases emissions of a car scrappage scheme in a time range of 24 years, from 1996 to 2020, reducing the average vehicle life from 12 to 10 years; the reference value of retirement age is calculated from the Italian vehicle fleet and used cars market statistics.

Two case studies are analyzed for two reference gasoline cars and one diesel car, representative of the European and USA fleet in 2005, considering emissions and consumption data respectively from the MIT and JRC reports published in 2008. In the baseline case it is considered the expected technological evolution of conventional gasoline vehicle both on propulsion and non-propulsion systems. The advanced case takes into account a higher technology improvement up to 2020, with hybrid propulsion systems.

Both case studies show that the GHG emissions on the whole vehicle life cycle are neutral with respect to car replacement acceleration, provided that the promoting incentives are restricted to the more energy efficient new vehicles, and new technologies development expected for 2020 will be achieved.

Keywords: GHG emissions, life cycle analysis, car scrappage, passenger cars.

WIT Transactions on Ecology and the Environment, Vol 131, © 2010 WIT Press
www.witpress.com, ISSN 1743-3541 (on-line)
doi:10.2495/EEIA100111

1 Introduction

Since the 1990s European and North American governments have adopted policies of scrappage subsidies in order to stimulate the early voluntary removal of used cars. Such grants may be used as a policy tool to reduce overall pollutant emissions from passenger cars by retiring the oldest vehicles, with the worst emission standards or no standard at all. Within the European Union, France, Greece, Hungary, Ireland, Italy, and Spain offered subsidies that required purchasing a new vehicle as a replacement. Instead, Denmark and Norway, similarly to the United States and Canada, offered subsidies just for scrapping old vehicles [1–3].

The scrappage incentives adopted in Italy in the last two years required the replacement of the old cars with new ones respecting standards even for greenhouse gas emissions. On the whole, those policies have been successful in increasing the rate of vehicle fleet replacement, diminishing national pollutant emissions.

A potential environmental trade-off of measures related to the reduction of the average life of a car, is the corresponding increasing rate of natural resources use, like materials and energy use in car manufacture and waste. This is the reason why in the last ten years different national and international research institutes have investigated the environmental effects of the scrappage policies from a life cycle point of view, under specific hypothesis of vehicles technical improvements in specific regional patterns.

The purpose of this study is to estimate the potential future impacts on GHG emissions of a scrappage scheme for passenger cars in 24 years, from 1996 to 2020, considering the technology evolution on the basis of the studies in literature.

The most recent and complete applications of Life Cycle Assessment (LCA) for the transport sector are presented briefly in Section 2. Section 3 illustrates methodology, assumptions and input data used in the present study. Finally, in Section 4, the results and conclusion are reported.

2 Life cycle analysis: literature review

The life cycle approach in assessing impacts associated with a product or service is a method that includes all the stages of a product system, from the extraction of raw materials, through the production of materials and intermediates, the product use or the service operation, to recycling and final disposal. In this way, the estimated impacts are prevented to be associated to wrong phases, helping policy makers, industry and all the stakeholders to improve or modify the product process and use in a way of reducing the more impacting phases.

Different life cycle applications were developed for addressing various life cycle-generated impacts and possible measures to minimize them: LCA, Life Cycle Management (LCM), Life Cycle Costing (LCC), etc. Some examples of those applications are: product development and improvement, process and service operation, strategic planning, technological impact assessment, public policy making, marketing [4].

For vehicles LCA, the following five processes have to be taken into account (Figure 1):

- car production: raw material extraction, material transformation and car assembly;
- replacement and spare parts production (tyres, battery, lubricants and refrigerants)
- car disposal and waste treatment (end-of-life – EOL);
- fuel transformation process upstream to fuel consumption (well-to-tank – WTT), composed of primary energy extraction, fuel production and distribution;
- vehicle use (tank-to-wheel – TTW).

In this study the first three processes are considered in one stage, called Materials; the WTT and TTW phases together represent the complete energy conversion chain, the so called well-to-wheel (WTW) [5, 6].

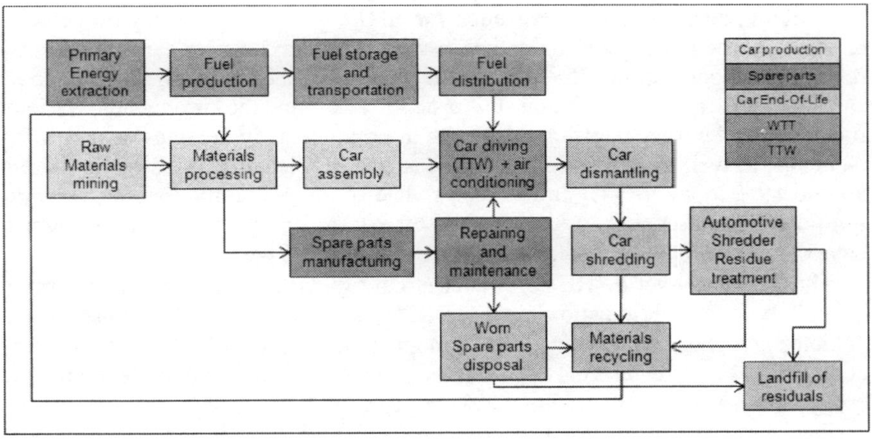

Figure 1: Car life cycle diagram.

The MIT studies were the first at focusing on the impacts of vehicle production and end-of-life [7]. In the 2008 a report commissioned by the European Commission was published by JRC with an extensive LCA study of current European passenger cars and forecast of vehicle technologies and alternative fuels evolution [6].

2.1 The MIT analysis

The objective of the MIT study was to assess and compare the potential reduction of fuel consumption and GHG emissions deriving from the evolution of light-duty vehicles technologies during the next 25 years, from a life cycle prospective.

Fuel consumption and GHG emissions of an average car, both currently and in the future, were quantified for a combination of the more implemented/promising technologies. All considered vehicles have the same performance and size; weight, battery and fuel cell type are those representative in 2005; as a matter of fact, MIT assumed that any weight reduction for future vehicles is realized using lightweight materials, that are more energy intensive than steel; similarly, the use of energy intensive materials for components such as batteries and fuel cell membranes in hybrid and fuel cell vehicles (FCVs), is expected to increase energy consumption and GHG emissions.

Consumption and emissions of different propulsion technologies in the vehicle use stage (TTW) were calculated with simulations over several standard driving cycles. The size and performance of future vehicles was set constant at the level of the Toyota Camry, with a 2.5-liter engine, used as a representative car. Scaling laws were used to estimate the evolution of individual vehicle components of the future vehicle. For the non-propulsion systems, the improvement in vehicle aerodynamics and tire rolling friction is assumed to hold a constant annual rate of reduction respectively of 1% and 1.65%. A 20% reduction in curb weight was assumed for all the future gasoline engine vehicles at constant size and safety. Reduction in vehicle size and weight can significantly reduce fuel consumption. Every 10% of weight reduced from the average new car can cut fuel consumption by around 7%. For the propulsion system, transmission improvements coupled with improvements in bearings, gear sealing elements, as well as hydraulics, can increase the efficiency of transmissions from around 89% today to 94% in the future. The future Naturally Aspirated Spark-Ignition (NA-SI) engines were projected by extrapolating historical trends which have demonstrated improvements on the order of 0.5% per year.

The gasoline hybrid (HEVs) model is configured as a single-motor parallel hybrid, with a hybridization ratio of 25%, sufficient to fulfill most of the vehicle's regenerative braking requirement under the USA's "typical" driving conditions. For the future HEV model, it was assumed that battery specific energy improves at a rate of about 2% per year.

For the fuel chain (WTT), the conventional gasoline production was considered equal to 24 MJ/MJ delivered in tank, with no efficiency improvements until 2035.

Figure 2 shows the MIT results for the GHG emissions for the three LC stages – WTT, TTW and Materials – for the average gasoline Spark-Ignition Engine vehicles (SIE) in 2005 and projected to 2035. Plug-in hybrid with 30-miles all electric range (PHEV-30), Hydrogen Fuel Cell Vehicle (FCV) and Battery Electric Vehicle (BEV) are also considered, denoting the uncertainty about the WTT GHG emissions for electricity generated from coal (upper bound) and natural gas (lower bound). All 2035 vehicles were supposed to have more efficient transmissions, 20% lower weight and reduced drag and tire resistances.

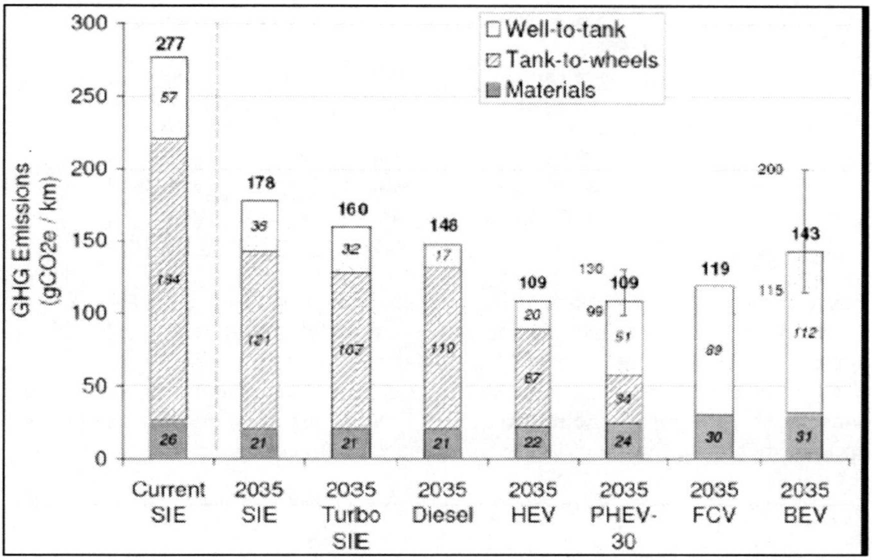

Figure 2: Lifecycle GHG emissions [7].

The study concludes that a 30–50% reduction in fuel consumption of new light-duty vehicles is feasible over the next 20–30 years. The greater uncertainty lies with the time necessary to achieve these changes, rather than the technological options available to realize them.

2.2 The JRC analysis

The JRC report provides a comprehensive analysis of the vehicle technical improvement options that could be achieved and marketed within 2020, and estimates the size of the environmental improvement potentials related to each option for each stage of the vehicle LC [6].

The technical improvements considered were compared to two reference car models, for gasoline and diesel cars, representative of the new passenger cars fleet in the EU-25 in 2005, averaging the characteristics values from the automotive market data (Table 1). The improvement options were evaluated including technical information from the industry, scientific publications and recent Commission's impact assessments – supporting policy developments. All the directives and proposals of the Commission were considered, regarding the new air pollution standards, for CO_2 emissions standards from passenger cars, for a new Directive regarding fuel quality, about vehicle end-of-life targets and the use of biofuels.

The life cycle data for the extraction, processing and production of materials were obtained from the Ecoinvent database and the International Iron and Steel Institute data, while the Energy used (and fuel mix) for the assembly phase was

Table 1: Main characteristics of the reference vehicles.

base case car	unit	Petrol	diesel
Average Lifespan	years	12.5	12.5
Emission standard		EURO4	EURO4
Average Annual distance	Km	16 900	19 100
Average total mileage	Km	211250	238750
Average Cylinder capacity	cm^3	1 585	1 905
Average Weight	Kg	1 240	1 463

derived from a study published by Volkswagen. The GHG emissions for the TTW phase were estimated using emissions database of real world driving cycles, especially the ARTEMIS database. For the WTW phase the JRC(IES)/CONCAWE/EUCAR study and the Ecoinvent database were considered. In Figure 3 the resulting GHG emissions of some of the improving options are shown for the gasoline and diesel passenger cars.

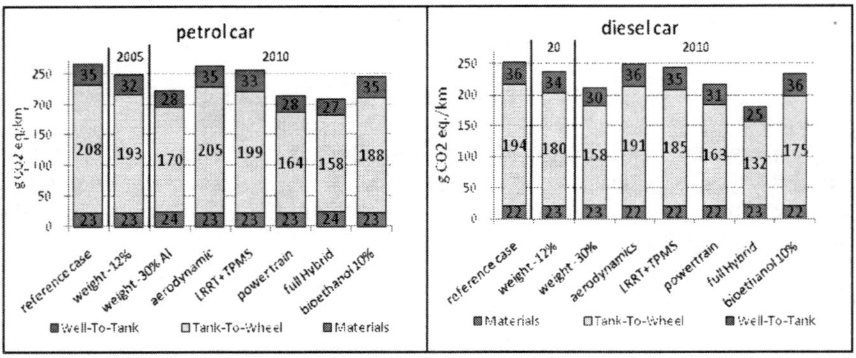

Figure 3: LC GHG emissions for technological options.

3 Methodology and data description

The data collected from the studies described in section 2 were utilized to compare two scenarios of different average car lifespan, considering the Italian car fleet characteristics:

- A "Business As Usual" scenario, with the average age of retired cars estimated from National database [8], equal to 12 years (Figure 4);
- An "accelerated scrappage" scenario, with the average age equal to 10 years. It was fixed on the basis of the used cars market data, and it is expected to be economically neutral for a car owner to sell it to a re-user or to scrappage if a 700 € public subsidy is forecast in the latter case [9].

The GHG emissions were estimated in a range of 24 years, from 1996 to 2020, considering 10 car owners, so to have 240 years of total lifespan for the 10 cars fleet, for both the following scenarios:

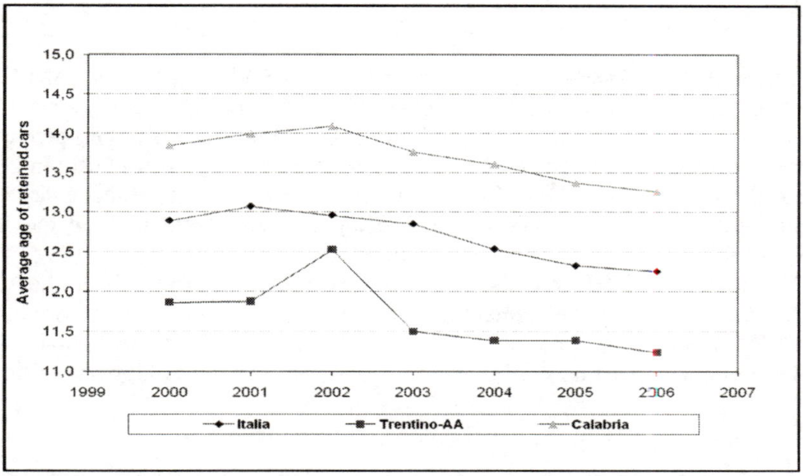

Figure 4: Italian retired cars average age.

- BAU scenario, where all the cars bought in 1996 are replaced after 12 years, first time in 2008;
- "Accelerated scrappage" scenario, where 6 owners continue to replace the car after 12 years, the other four replace the car every 8 years, in 2004 and in 2012. The resulting average car lifespan is 10 years.

Two scenarios of technological improvements were evaluated: a baseline scenario, with a trendy development of the conventional vehicles, and an advanced option, with enhanced hybrid vehicles. North-American cases, based on MIT data, and European cases, based on JRC results, were analyzed. All the emissions were calculated applying the scaling methods for the improvements.

3.1 American case studies

The two technological scenarios considered correspond to the 2035 SIE car and the 2035 HEV car (Figure 2). The data used for the GHG emissions calculations of the two scenarios are reported in Table 2.

Table 2: GHG emission for the 3 phases.

LC phase	GHG emissions	1996	2004	2008	2012	2020
Materials GHG metric tons/vei.	Baseline	7,7	7,7	7,9	8,1	8,5
	Advanced	7,7	7,7	8,1	8,5	9,3
WTT g CO_2 eq./km	Baseline	62	57	54	52	46
	Advanced	62	57	52	47	38
TTW g CO_2 eq./km	Baseline	210	194	185	175	156
	Advanced	210	194	180	166	139

3.2 JRC case studies

Three technological scenarios (the baseline one, two advanced) were calculated for both gasoline and diesel cars, considering the ACEA statistics on CO_2 emission of sold cars for the 1996, and the TTW CO_2 emissions targets proposed by the European Commissions, 120 gCO_2/km as an average of sold cars in 2020 in a softer hypothesis and 95 gCO_2/km in an harder one. To meet those targets a technologies combination was assumed, using data and evaluations of the JRC report (Table 3).

The total TTW GHG emissions were established starting from the CO_2 emissions, estimating the percentage of CH_4 and N_2O emission through the COPERT 4 emissions for EURO 2 cars (1996) and EURO 4 cars (2004) [10], using the Italian speed and mileage percentages of urban-rural-highways driving

Table 3: Technologies combinations.

	gasoline car			**diesel car**		
TTW 2005 g CO_2/km	169			155		
technologies scenario	**baseline**	**advanced**	**Adv. hybrid**	**baseline**	**advanced**	**Adv. hybrid**
Target gCO_2/km	120	95	≤95	120	95	≤95
technologies	Aerody-namics tyre powertrain	weight-12% aerodyn. tyre powertrain	full hybrid aerodyn. tyre powertrain	Aerodyn. tyre powertrain	weight-12% aerodyn. tyre powertrain	full hybrid aerodyn. tyre powertrain
Em. Reduction	73%	59%	49%	78%	64%	47%
2020 em. g CO_2/km	123	100	83	121	100	72

Table 4: GHG emissions for the 3 phases.

GHG em. of LC phases		**gasoline car**					**diesel car**				
LC phase	**Tecnologies scenario**	**'96**	**'04**	**'08**	**'12**	**'20**	**'96**	**'04**	**'08**	**'12**	**'20**
Materials t CO_2 eq.	Baseline	4,70	4,70	4,70	4,70	4,70	5,20	5,20	5,20	5,20	5,20
	Advanced	4,70	4,70	4,70	4,70	4,70	5,20	5,20	5,20	5,20	5,20
	Adv. hybrid	4,70	4,70	4,76	4,82	4,94	5,20	5,20	5,27	5,33	5,46
TTW g CO_2 eq./km	Baseline	189	173	162	150	126	176	158	149	139	121
	Advanced	189	173	153	134	100	176	158	143	129	100
	Adv. hybrid	189	173	149	126	83	176	158	136	115	72
WTT g CO_2 eq./km	Baseline	32	29	27	24	19	30	27	25	22	18
	Advanced	32	29	25	21	15	30	27	24	21	15
	Adv. hybrid	32	29	25	0	12	30	27	22	18	11

modes. Future vehicles were supposed to have no significant GHG emissions either then CO_2. The JRC average total mileages for both petrol and diesel car were used.

The WTT emissions were calculated considering an efficiency improvement from 17% in 2000 to 15% in 2020 on the fuel consumed in the fuel cycle, as resulted in the JRC-CONCAWE [11]. The Materials processes emissions were estimated considering the same emissions from 1996 to 2004, and applying the improvement percentages on the future vehicle productions as indicated in the JRC work. The resulting values are shown in Table 4.

4 Results and conclusions

Within the hypothesis adopted in this study, both the American and the European accelerated scrappage scenario produce an increase in GHG emissions around 1%, as shown in Table 5, that can be considered within the calculations errors.

Table 5: GHG emissions for all the case studies.

case studies	**American**		**European petrol car**			**European diesel car**		
scenario GHG t CO_2 eq.	baseline	advanced	baseline	advanced	advanced hybrid	baseline	advanced	advanced hybrid
BAU	1382	1369	917	903	893	972	957	940
Accelerated Scrappage	1403	1388	926	910	899	983	966	947
% GHG reduction	**-1,5%**	**-1,4%**	**-1,0%**	**-0,8%**	**-0,7%**	**-1,1%**	**-0,9%**	**-0,7%**

The discrepancies between the two case studies are related to the car characteristics and annual mileages: a 2.5-liter engine and 20000 km/year for the American vehicle, a 1.6-liter engine and 16900 km/year for the European petrol car. Moreover, the WTT phase for American case is more energy consuming than the European one.

In Figure 5 the detailed emissions for the three LC phases are shown for all the case studies. It results that the reduction in GHG emissions obtained during vehicles operation is lost with the increase in the vehicle production stage.

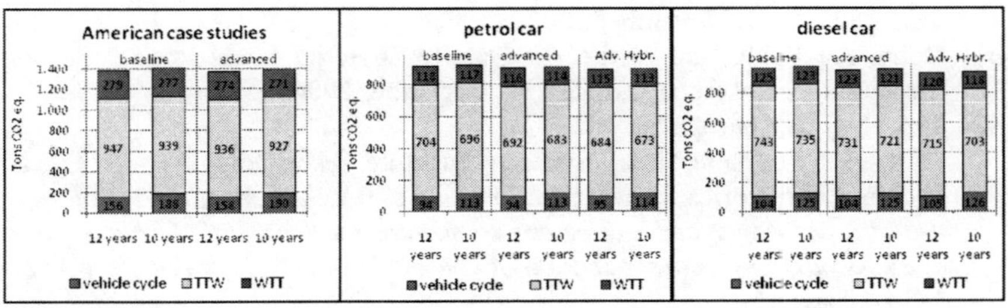

Figure 5: GHG emissions.

In conclusion, the considered scrappage policy results neutral from the GHG emissions point of view, if targeted to a replacement with conventional cars with more stringent CO_2 emissions standard. For technology advanced scenarios, the gaps between BAU and accelerated scrappage are even smaller. Nevertheless, it is to emphasize that this estimation is very cautious, due to the lack of data about the batteries end-of-life and re-use.

Table 6: Technologies impacts for the petrol car [6].

Impacts normalized to a 100 km distance for petrol car (%)	Reference	Aerodynamics	Tyres	Power train improvements	Weight reduct. -12%	Hybrid car
abiotic depletion	100	100	100	100	98,4	55,1
climate change	100	98,7	96	80,5	93,9	78,4
ozone depletion	100	98,6	95,9	79,7	93,7	77,2
photochemical oxidation	100	99,1	97,3	86,6	95,7	75
acidification	100	99	97	85,2	96,3	90,6
eutrophication	100	99,1	97,4	87,2	96,7	83,1
2.5 microns particulate matter	100	98,9	96,8	84,3	100,8	88
consumption of primary energy resources	100	98,7	96,1	80,9	94,3	78,6
solid waste	100	99,6	98,9	94,6	103,3	104,3
Monetarised Aggegated impacts		98,5	96	81,5	94,4	83,1

Considering also other environmental impacts, as those listed in Table 6, by using the JRC data, the scrappage subsidies are globally efficient, at least so far as improvement in vehicles emission standards are provided thanks to UE legislation (see Table 6 for the petrol car case).

References

[1] ECMT, *Conclusions and recommendations on scrappage schemes and their role in improving the environmental performance of the car fleet* CEMT/CM(99)26/FINAL. European Conference of Ministers of Transport, June 1999

[2] Commission of the European Communities, Commission staff working document accompanying the Communication From The Commission – *A European Strategic Energy Technology Plan (SET-Plan) – Capacities map*, SEC(2007)1511, November 2007

[3] Defra, An Evaluation of the Air Quality Strategy: Local Road Transport Measures – Final Report, ED50232, December 2004

[4] http://lct.jrc.ec.europa.eu/

[5] Michael Spielmann, Paul Scherrer Institute, Hans-Jörg Althaus, Swiss Federal Laboratories for Materials Testing and Research *"Can a prolonged use of a passenger car reduce environmental burdens? Life Cycle analysis of Swiss passenger cars"*, Journal of Cleaner Production 15 (2007), p.1122-1134, Elsevier

[6] F. Nemry, G. Leduc, I. Mongelli, A. Uihlein , *Environmental Improvement of Passenger Cars (IMPRO-car)*, JRC 40598, EUR 23038 EN, March 2008
[7] Anup Bandivadekar et al., *"On the Road in 2035: Reducing Transportation's Petroleum Consumption and GHG Emissions"*, MIT Report No. LFEE 2008-05 RP, July 2008
[8] Automobile Club d'Italia, "Annuario Statistico 2007", 2008
[9] Quattroruote, Gennaio 2009
[10] http://lat.eng.auth.gr/copert/
[11] Edwards R, Larivé JF, Mahieu V, Rouveirolles P., *Well-to-Wheels analysis of future automotive fuels and powertrains in the European context,* European Commission DG JRC/IES, CONCAWE and EUCAR, 2006

Section 3
Decision support systems

Towards a decision support tool: sensitivity mapping of the French Mediterranean coastal environment (a case study of fishery and lodging)

C. Scheurle, H. Thébault & C. Duffa
LERCM, IRSN, La Seyne-sur-mer, France

Abstract

Coastal management implicates decision making processes in order to meet conflicting interests, to target a sustainable development and to guarantee efficient actions to face man-driven pressures. Therefore, adequate decision support tools are needed. The present study is embedded in a project that has the objective to develop an operational simulation tool particularly adapted to manage a crisis situation provoked by accidental marine pollution on the French Mediterranean coast. So as to deal with such a risk, our study is intended to contribute to the analysis of the coastal vulnerability. The vulnerability depends on the exposure to risk factors often linked to human activities, as well as on the environmental and socio-economic characteristics. It is intended to determine these characteristics, to estimate their "values" and to assess the sensitivity of the coastal environment, while the overall aim is to anticipate possible impacts. As demonstrated in various past cases, e.g. an accidental hydrocarbon release impacts different ecosystem types and the environmental services they deliver and that sustain considerable socio-economic activities, such as fishery and tourism. The sensitivity to such impacts has been evaluated for the French Mediterranean coastal environment and sensitivity indices have been attributed to defined coastal zones. Taking seasonal variations into account, here we present sensitivity maps of two "emblematic" coastal activities that are directly or indirectly linked to seawater quality, fishery and lodging.

Keywords: coastal management, evaluation, environmental services, market and non market goods, marine pollution.

WIT Transactions on Ecology and the Environment, Vol 131, © 2010 WIT Press
www.witpress.com, ISSN 1743-3541 (on-line)
doi:10.2495/EEIA100121

1 Introduction

1.1 Development of a decision support tool within the context of coastal management

Coastal management addresses the interactions of natural ecosystems and human activities. It implicates decision making processes and conservation strategies trying to balance environmental values and anthropogenic uses. In order to meet conflicting interests, to target a sustainable development and to guarantee efficient actions to face man-driven pressures, adequate decision support tools are needed that facilitate an application of mechanisms, monitoring, protection and alerts.

In order to assist management requests, in particular in the case of accidental marine pollution, the project CLARA 2 ("Calculs Liés Aux Rejets Accidentels en Méditerranée", 2006-2010) aims at providing a specific decision support tool for the French Mediterranean coast. The French National Research Agency ("Agence National de la Recherche", ANR) supports the project within its programme on eco-technologies and sustainable development ("Programme Ecotechnologies et Développement Durable", PRECODD). Twelve partners (a consortium of academic research laboratories, industries, civil services and SME) are involved in the project, which is labelled by the French cluster of risk management and vulnerability of the territory ("Pôle Gestion Risque et Vulnérabilité de Territoires").

By developing an operational simulation tool, CLARA 2 has the objective to assist in the diagnosis of an accidental discharge of chemical substances (e.g. hydrocarbons) in the coastal environment. To face such pollution and its consequences the assessment of risks, the modelling and forecasting of pathways of the contaminant and the evolution of its concentration in the air-sea system are needed. Moreover, specifically adapted information on the coastal environment and use is necessary in order determine particularly sensitive zones and to propose recommendations to methods, means and priorities of interventions.

1.2 Assessment of risk and sensitivity to determine the vulnerability of the coastal environment

Risk factors are often linked to maritime traffic. Today, about 30% of the international maritime traffic concerns Mediterranean harbours or are on transit and estimated 50% of the transported merchandises are supposed to present a risk (Aprin et al. [1]). In the Mediterranean, the sector of maritime shipping develops and port facilities, tonnage and frequency of traffic steadily augment. Transport of hydrocarbons takes an important part. The accidental release of hydrocarbons, or more in general accidental marine pollution, is considered as one of the major risks linked to human activity impacting the coastal environment.

However, the impacts on the coast not only depend on the type of substance generating the pollution and its extension but also on the affected zone itself. It depends on its environmental (physical, dynamical and biological) and socio-

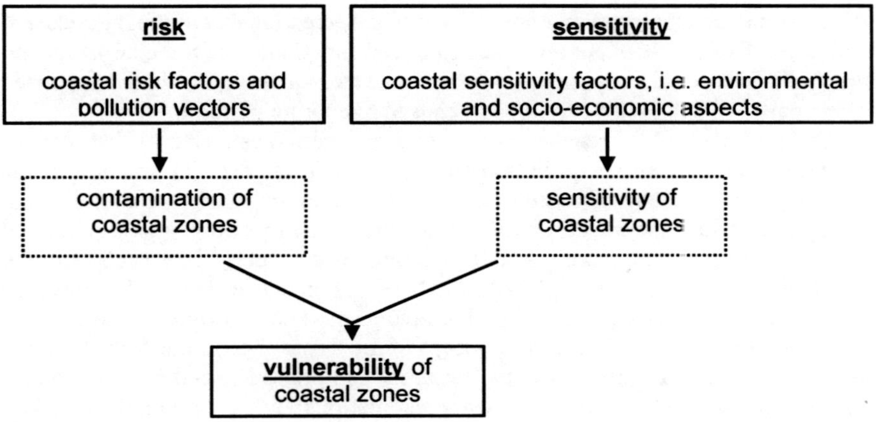

Figure 1: The vulnerability of coastal zones depends on the sensitivity of its environmental and socio-economic factors, as well as on its exposure to risk factors (Duffa et al. [2]; adapted by C. Scheurle).

economic aspects. These aspects characterize a coastal zone, and they indicate and describe its sensitivity (sensitivity factors; Figs. 1 and 2). By evaluating them, sensitivities of the coastal environments to accidental marine pollution are here assessed for the French Mediterranean coast. Subsequently their geographic distribution can be determined which allows the "sensitivity mapping".

Coupling risk and sensitivity assessments then allows determining the coastal zones vulnerability (Fig. 1). Within the context of coastal management the analysis of the vulnerability is important in order to support decisions and to deal with the risks of marine pollution.

1.3 Approach "services provided by the coastal environment" – placing the present study within its context

In order to contribute to management and decision making processes as well as conservation strategies, approaches are made to determine "values" of marine and coastal ecosystems. They deliver a wide range of environmental services (cf. e.g. Beaumont et al. [3]) that have important socio-cultural impacts and that sustain considerable economic activities such as fishing and tourism.

Within this context, we aim at estimating values (monetary or non monetary) for the French Mediterranean coast. Taking environmental and socio-economic aspects into account, the undertaken evaluation of quantitative and semi-quantitative information is then used to determine respective sensitivities assuming that a higher "value" indicates higher sensitivity.

Three main sensitivity factors have been identified: (a) physical and dynamical, (b) biological and (c) socio-economical. Amongst the factors (a) and (b), we consider physical characteristics such as the coastal geomorphology and the type of sediments as well as the marine biodiversity (e.g. benthic assemblages and rare species). The socio-economic part (c) includes market (e.g.

fisheries and aquaculture) and non market (e.g. recreational activities) goods and activities. The activities are all linked to a well functioning of the ecosystem and are indirectly or directly depending on seawater quality; the latter supposed being particularly sensitive to impacts caused by marine pollution.

Our inventory of 26 relevant socio-economic activities (criteria) that may be affected in case of marine pollution covers six main sectors: (1) products of the sea (e.g. fisheries and aquaculture), (2) extraction of marine materials (e.g. extraction of salt; therapeutic uses), (3) coastal tourism, sports and recreational activities (e.g. lodging, cruises and yachting, water sports like scuba diving, beaching, recreational fishing), (4) efforts of maintenance, (5) marine research and educational activities, (6) cultural attractiveness and maritime heritage.

Within the goods and services concept of the "Total Economic Value" (TEV; cf. e.g. Ledoux and Turner [4]) this inventory comprises use and non use values. The economic analyses (i.e. monetary estimations) however mainly consider direct uses. Here, we present the case study of two "emblematic" coastal activities, fishery and lodging.

Our analyses position in an *ex ante* case study, i.e. the evaluation takes socio-economic activities into account that may potentially be impacted in this particular Mediterranean coastal region. The dimensions of impacts and losses that an accidental release of hydrocarbons may cause have been demonstrated in various past cases (*ex post* analyses, e.g. Garza-Gil et al. [5], Loureiro et al. [6]).

1.4 Resume of the present study's main objectives

The project CLARA 2 aims at developing a support tool for coastal management and decision making in case of accidental marine pollution at the French Mediterranean coast. As part of this project, our work contributes by characterising defined coastal zones in terms of their environmental and socio-economic aspects.

The present case study focuses on socio-economic aspects, namely the monetary evaluation of two selected activities, fishery (directly linked to seawater quality) and lodging (indirectly linked to seawater quality). This evaluation helps to determine sensitivities which are "translated" into indices. The sensitivity indices are displayed in a geographical context, on sensitivity maps. The maps are supposed to show a relative importance of these exemplary activities within the coastal zones intending to identify "hot spots" of activity and thus action needed.

2 Elaboration of sensitivity maps of the French Mediterranean coast

2.1 Study area and temporal frame

The study area is situated in the north-western part of the Mediterranean. It embraces the whole French Mediterranean coast which represents ca. 1800 kilometres of coastline.

From an administrative point of view, the study area is structured in three regions that are subdivided into departments (the latter in brackets; from west to east): Languedoc-Roussillon (Pyrénées-Orientales, Aude, Herault), Provence-Alpes-Côte d'Azur (PACA; Gard, Bouches-du-Rhône, Var, Alpes-Maritimes) and Corsica (Haute Corse, Corse Sud). A total number of 189 French municipalities border the Mediterranean Sea of which some are incorporated in agglomerations like Marseille (Bouches-du-Rhône), Toulon (Var) and Nice (Alpes-Maritimes).

In order to elaborate the respective sensitivity maps, the study area has been divided in zones. The division in zones is based on the so-called SDAGE ("Schéma Directeur d'Aménagement et de Gestion des Eaux"), the French outline for the organization of the development and management of water resources. Taking into account main environmental characteristics (e.g. hydrographical dynamics) as well as requirements to management and decision making processes, 32 zones have been distinguished. Seawards the outermost limit of either four miles and/or 50 meters depth has been chosen for the coastal zones.

Tourist activities and their impact mainly during the summer season make a seasonal differentiation necessary (summer: April to September, winter: October to March). The scenario of pollution is limited to a short term event. I.e. the impacting event is supposed to last no longer than days up to, at maximum, a few months.

2.2 Evaluation method, data collection and treatment

In order to elaborate sensitivity maps of the coastal zones facing the French Mediterranean and to assess the sensitivity of each zone in case of an accidental marine pollution, detailed information on (environmental and) socio-economic aspects is needed.

Concerning the socio-economic aspects, most recent data accessible on 26 criteria (cf. 1.3) has been collected. Then it has been compiled to match to the 32coastal zone units (cf. 2.1). After collection of data of socio-economic factors and criteria together with their seasonality, and after an evaluation process, sensitivity indices have been attributed (Fig. 2) to the seaward side, the "marine sector". These indices "translate" the sensitivity criteria on a scale from 1 (less sensitive) to 5 (very sensitive).

The next step will comprise the elaboration of a "global" index, encompassing all socio-economic criteria. Since summer and winter season are regarded separately, two sets of maps result. The produced sensitivity maps will then be compiled in an atlas that will be implemented into the operational simulation tool (cf. Scheurle et al. [7]).

In the context of the present study, we follow this approach for two criteria and focus on two selected socio-economic activities, fishery and lodging. These stand exemplary for activities that are directly or indirectly depending on seawater quality. The catch of seafood, for example, may directly be impacted in case of marine pollution, whereas an onshore tourist activity like the hotel business is rather indirectly dependant on the quality of seawater.

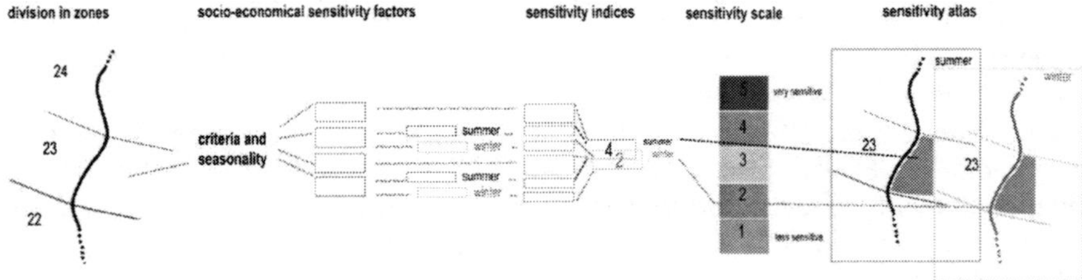

Figure 2: Scheme of the elaboration of a socio-economic sensitivity map. After the division in zones, the main following steps include data collection relative to various criteria representing the sensitivity factors, their evaluation and the attribution of indices, as well as the representation for each of the zones (hypothetic example) (Scheurle et al. [7], adapted by C. Scheurle).

2.2.1 Data on an activity directly linked to seawater quality – the example of fisheries

The coastal fishery comprises several different techniques quite specific in each regional context. The intensity of fishery activity within each of the coastal zones, as well as the active fishermen, have been taken into account so as from case to case, average market prices of seafood. Partly data on the revenue were accessible and have been used for the present analysis. No homogenous dataset was available and therefore various sources have been inquired (e.g. IFREMER/SIH Méditerranée [8], OFIMER [9] and pers. comm. "Affaires Maritimes en Méditerranée - DRAM Corse/DDAM/DIR Service économie").

To briefly summarize the evaluation that results in a monetary estimation (in Euros): for Languedoc-Roussillon and PACA the fishery activity per coastal zone has been estimated (month-vessel); the revenues per characteristic fishery type have been determined for each region and redistributed in relation to the fishery activity per coastal zone; presumably 75% of the estimated revenues are due to summer activity, 25% concern the winter season. The same percentage has been used for the Corsica region. However, in this case, the revenues per fisherman (anonymous) for each zone have been accounted.

The evaluation results were then used to attribute sensitivity indices on the 1 to 5 scale (1, less sensitive, summer: < 300 K€, winter: < 250 K€; 5, very sensitive, summer: > 8.000 K€, winter: > 2.500 K€; see above; only summer classification is displayed in Fig. 3) for fishery activity in each of the coastal zones.

2.2.2 Data on an activity indirectly linked to seawater quality – the example of lodging

The activity is basically linked to coastal tourism and comprises the hotel business, secondary residences and camping. Primarily data of the French National Institute of Statistics and Economic Studies ("Institut National de la Statistique et des Etudes Economiques", INSEE) have been used, completed by information obtained from local and regional agencies working on the analysis of the tourism sector.

To briefly summarize the evaluation process: the total numbers of each accommodation type on local (municipality) level have been used to calculate the capacity of accommodation for each coastal zone as well as their occupation (monthly; in percentages). Overnight stays and average expenses per day have been consulted. The summing up of the six summer and winter months results at first for each accommodation type, then for the whole lodging sector, in an evaluation (monetary; in Euros) for each coastal zone.

The resulting values were then used in order to attribute sensitivity indices on the 1 to 5 scale (see above). The lower class (1, less sensitive) integrates values from < 25.000 K€ for the summer season and values < 5.000 K€ for the winter season. The classification in the uppermost class (5, very sensitive) includes summer values of > 250.000 K€ and winter values of > 100.000 K€ (Fig. 4).

3 Sensitivity mapping of the French Mediterranean coast

3.1 Sensitivity maps of two selected socio-economic activities – results

The (here: monetary) valuation, as described above, allows to attribute sensitivity indices for each coastal zone and subsequently to present them cartographically. The "sensitivity mapping" results for two selected socio-

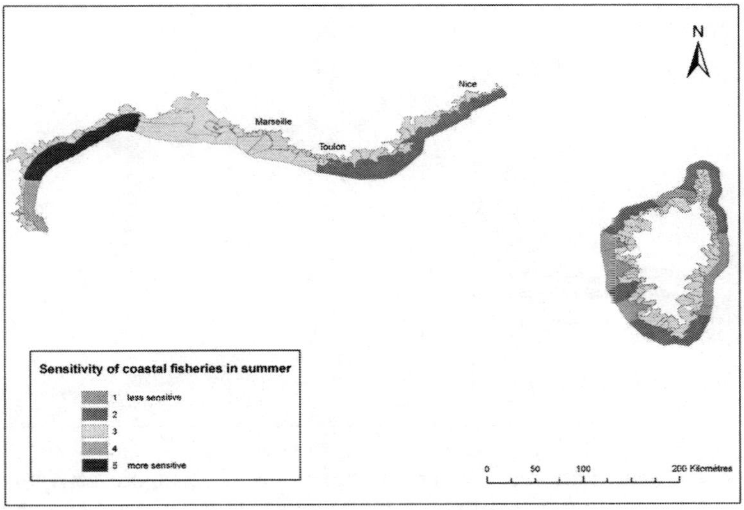

Figure 3: Sensitivity map of an exemplary socio-economic activity that is directly linked to seawater quality – fishery. The map shows the sensitivity indices attributed to fishery activity along the French Mediterranean coast during the summer season for each of the 32 coastal zones. The municipalities belonging to these defined zones are indicated in grey-shading on the landward side.

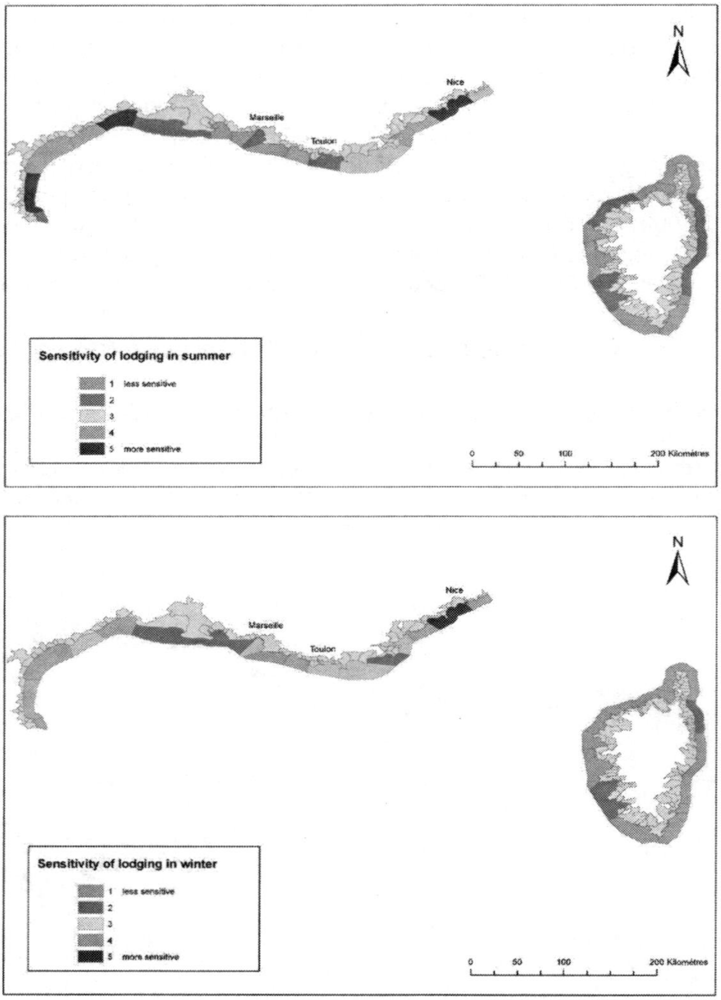

Figure 4: Sensitivity maps of an exemplary socio-economic activity that is indirectly linked to seawater quality – lodging. The maps show the sensitivity indices attributed to accommodation facilities along the French Mediterranean coast during summer (upper map) and winter (lower map) season for each of the 32 coastal zones. The municipalities belonging to these defined zones are indicated in grey shading on the landward side.

economic activities, fishery and lodging, are presented here. Both activities are linked to marine services and the one is directly linked to seawater quality the other one indirectly.

3.1.1 Sensitivity map of fishery activity (directly linked to seawater quality)

All along the French Mediterranean coast, fishery is considered as one of the important activities in terms of socio-economic aspects. Largely more intense fishery activity is to be found in the Languedoc-Roussillon region (cf. 2.1). High sensitivities linked to fishery activity therefore characterize the western French Mediterranean coast (Fig. 3; only summer season displayed). The adjacent eastward coastal zones of the PACA region (cf. 2.1 and Fig. 3) are as well ranked on the upper sensitivity scale, whereas Corsica seems to be comparatively less sensitive in case of an accidental marine pollution.

3.1.2 Sensitivity map of lodging (indirectly linked to seawater quality)

In particular due to the tourist activity at the French Mediterranean coast, the values (in Euros) on which the sensitivity indices are based on are relatively high. In summer, Languedoc-Roussillon (cf. 2.1 and Fig. 4) is the region that evidences high sensitivities. Ranked as particularly sensitive are also the coastal zones within the PACA region (cf. 2.1 and Fig. 4) with the agglomeration Nice as "hot spot". In winter, though ranked slightly less sensitive, the same coastal zones than in summer are rather more sensitive in case of an accidental marine pollution (Fig. 4). Then, however, the eastward continuation of the agglomeration Nice becomes relatively more sensitive due to its specific winter tourism.

3.2 Sensitivity maps of two selected socio-economic activities – discussion

The presented maps, elaborated on the basis of a monetary evaluation, allow characterizing the French Mediterranean coast with respect to the two selected socio-economic activities. With respect to an activity that is directly (indirectly) linked to seawater quality the relative sensitivities to an accidental marine pollution from one coastal zone to another can be distinguished.

Within this context, and outlining only very briefly here, three aspects need particular consideration. Firstly, the evaluation results are as good as the data available (the datasets were not homogenous i.e. due to the different sources). Secondly, they represent most likely an underestimation. Thirdly, even when displaying the sensitivities in the same manner, the range of values that stands behind the sensitivity scale is different for the two activities. These facts are important for the further use of the maps (cf. below) as well for their interpretation.

4 Conclusions and perspectives

The present study is embedded in a project which means to develop a decision support tool, particularly adapted to manage a crisis situation provoked by accidental marine pollution on the French Mediterranean coast. In order to contribute to the vulnerability analysis, our work aims at characterizing coastal zones and to assess relevant environmental and socio-economic aspects. An evaluation (monetary and/or non monetary) of these aspects helps to attribute respective sensitivities (sensitivity indices) and to put them in their geographical context.

On the example of selected socio-economic activities, here we present first results of sensitivity mapping of coastal zones along the French Mediterranean coast. To illustrate two socio-economically important sectors, we chose fishery and lodging as exemplary activities.

Fishery is an activity that is directly linked to seawater quality and therefore strongly depending on the well functioning of the marine ecosystems. By estimating monetary values linked to this activity, our study attempts an economic evaluation of one of the goods provided by the marine environment and "directly used and consumed" at a specific geographical scale.

Accommodation capacity and its occupation are linked to seaside tourism and largely profiting from the wide range of environmental assets; they in turn are indirectly depending on the sea and its water quality. By estimating its monetary values, our study therefore tries to partly economically evaluate services provided by the sea that are not directly consumptive.

Both exemplary evaluations differentiate between summer and winter season. On the basis of the seasonal evaluations, sensitivities have been determined for the coastal zones. The resulting indices, hierarchical and used like a cartographic measure, are displayed in their geographical context of the French Mediterranean coast.

In the course of the study, the next steps will envisage to combine the classification results presented and the various other socio-economic sensitivity maps in order to obtain one map representing all different aspects. This will possibly be accompanied by a multi-criteria analysis together with analysis on weighting factors. Given e.g. sanitary aspects, it is to discuss whether activities that are directly linked to seawater/environmental quality should be regarded as "more important" in the present context than indirectly linked activities.

Subsequently, it is planned to gather the socio-economic sensitivity results with the environmental sensitivity maps in a sensitivity atlas (Scheurle et al. [7]) and to integrate them in the development of the decision support tool.

The presented approach is considered as dynamic allowing improvements at any time. Moreover, it is considered as "generic" and therefore supposed to be adaptable for other coastal regions as well as adjustable in terms of its temporal context.

References

[1] Aprin, L., Lefloch, S., Garreau, P., James, A., Daniel, P., Daumail, V., Sanchez, C., Mouries, M., Casselman, C., Thébault, H., Etasse, C., Roure, J.F., Mercantini, J.M. & Olier, R., Développement d'un outil informatique opérationnel d'aide à la décision et de modélisation des pollutions en Méditerranée. *16ème Congrès de Maîtrise des Risques et de Sûreté de*

Fonctionnement, Avignon 6-10 October 2008, communication 1Z-8, 8p, 2008.

[2] Duffa, C., Mercat-Rommens, C. & Thébault, H., Radioecological sensitivity project on the French Mediterranean coastal environment. CIESM, *Rapp. Comm. Int. Mer Médit.*, **38**, p. 666, 2007.

[3] Beaumont, N.J., Austen, M.C., Atkins, J.P., Burden, D., Degraer, Dentinho, T.P., Derous, S., Holm, P., Horton, T., van Ierland, E., Marboe, A.H., Starkey, D.J., Townsend, M. & Zarzycki, T., Identification, definition and quantification of goods and services provided by marine biodiversity: Implications for the ecosystem approach. *Marine Pollution Bulletin*, **54**, pp. 253-265, 2007.

[4] Ledoux, L. & Turner, R.K., Valuing ocean and coastal resources: a review of practical examples and issues for further action. *Ocean & Coastal Management*, **45**, pp. 583-616, 2002.

[5] Garza-Gil, M.D., Prada-Blanco, A. & Vásquez-Rodríguez, M.X., Estimating the short-term economic damages from the Prestige oil spill in the Galician fisheries and tourism. *Ecological Economics*, **58**, pp. 842– 849, 2006.

[6] Loureiro, M.L., Ribas, A., López, E. & Ojea, E., Estimated costs and admissible claims linked to the *Prestige* oil spill. *Ecological Economics*, **59**, pp. 48-63, 2006.

[7] Scheurle, C., Thébault, H., Duffa, C. & Ami, D., Sensitivity Atlas of the French Mediterranean Coast. Özhan, E. (Ed.), *Proceedings of the Ninth international Conference on the Mediterranean Coastal Environment*, MEDCOAST 09, 10-14 November 2009, Sochi, Russia, MEDCOAST, Middle East Technical University, Ankara, Turkey, **1**, pp. 543-553, 2009.

[8] IFREMER, Système d'Informations Halieutiques (IFREMER/SIH) Méditerranée, www.ifremer.fr/sih, 2001 and 2009.

[9] Ministère de l'Agriculture et de la Pêche & Office National Interprofessionnel des Produits de la Mer et de l'Aquaculture (OFIMER), Bilan annuel de production des pêches et de l'aquaculture, 85p., 2007.

Funding evaluation model for the implementation of wastewater treatment projects through public private partnerships

A. Ch. Karmperis[1], A. Sotirchos[2], K. Aravossis[2] & I. Tatsiopoulos[2]
[1]*Directorate of European and Developmental Projects Management, Ministry of Defence, Greece*
[2]*National Technical University of Athens, Greece*

Abstract

This paper examines the assessment and funding rate estimation process for the implementation of wastewater treatment (WT) projects through Public Private Partnerships (PPPs). The study, having as strong theoretical foundation the Cost Benefit Analysis (CBA) methodology and the quantitative Value for Money (VfM) assessment process, develops a new algorithmic type model, in order to present a process for the funding evaluation of PPP type WT project's initial investment. The model applies in those projects that are considered to be financed by both public and private sectors and further could be potentially co-financed by the European community. The model is tested in a WT project case study and calculates the upper and lower limits of the public and private sectors' funding rates in the initial investment. Due to the fact that a PPP is not a solution option but may be the procurement choice for a preferred solution option, the new model can be a useful tool to project examiners during the feasibility stage of WT projects, in order to evaluate alternative funding scenarios and propose the most suitable in each case option to decision-makers.
Keywords: wastewater treatment projects, public private partnerships, model, evaluation, cost benefit analysis, value for money, funding scenarios.

1 Introduction

Public Private Partnerships (PPPs) are contract types that have been used worldwide by the majority of countries over recent decades [1, 2]. Reeves [3],

WIT Transactions on Ecology and the Environment, Vol 131, © 2010 WIT Press
www.witpress.com, ISSN 1743-3541 (on-line)
doi:10.2495/EEIA100131

underlines that there is no common definition of PPPs, so definitions given by many authors show little differences [4, 5]. However, a common point between different approaches is that PPP contracts are long-term agreements for cooperation between the public and private sector to provide high quality infrastructure, products or services, delivered via a process of applied risk-sharing, resources and profits, while the duration of a PPP project's operational phase is 10 to 30 years [6]. However, an undeniable fact is that PPP projects have been rapidly increased over the last decades. For the period 1990 to 2007, the World Bank Group's database [7] shows that the peak of investment projects in the water and sewerage sector was in 1997. Additionally, in 2007, investments in the specific sector amounted to US$ 3 billion, within the US$ 2 - 3 billion range of the three previous years. For the period 1985 to 2004, water and sewerage sector projects through PPPs in the United States were 41% out of 364 in total and cost US$ 82 billion [8]. In Central-East Asia, large water projects were implemented through PPPs [9]. Specifically, WT projects have been executed in Shanghai and Chengdu, China, in Ahmedabad and Chennai, India, in Surabaya, Indonesia, in Bangkok, Thailand and in Ho Chi Minh, Vietnam, [10, 11], mainly following the Build Operate Transfer (BOT), which is the most used type in PPPs [12]. Moreover, for WT projects implementation, BOT contracts have been also used in Izmit, Turkey, in Chihuahua, Mexico, in Johor, Malaysia, a Build Own Operate contract in Sydney, Australia, full privatization contracts in England and Wales [13]. In the European Union (EU) also, PPP markets are continuously growing according to each country's model. This fact led the Commission to publish the guidelines for PPPs [14], as well as the green book on PPPs [15], which is a book of 22 questions, in order to collect data from member states and establish a common legislative framework. Conclusively, due to fiscal limitations that the global economic crisis induces, it is expected that PPP contracts will continue to play an important role in future public procurements.

2 Evaluation of PPP projects

In the literature, several methods for the financial evaluation of PPP projects are suggested, including the Cost Benefit Analysis (CBA) [16] and the Public Sector Comparator (PSC) [17]. Generally, it is recommended that all public procurements of goods and services should be based on the best Value for Money (VfM), which is defined as the optimum combination of whole life cost and quality, in order to meet the requirement [18]. According to Grimsey and Lewis [19], main alternative approaches that can be distinguished, are the full CBA of the public and private options, the PSC before bids invited and the United Kingdom (UK) style of VfM assessment. Particularly, in the UK as well as in Australia, the qualitative and quantitative processes that are used to determine the VfM, take into consideration all the costs and benefits included in a project's lifecycle [20]. Moreover, in the UK Private Finance Initiative (PFI) projects, incorporated in PPP programmes, were also developed. These kinds of projects, which started in 1992, had doubled by 1994, reached almost 500 in

2002 [21] and there were approximately 668 by the end of 2009 [22]. PFI are based in public services provision, operated and also financed by the private sector [23, 24]. Currently, the existing UK guide for the VfM assessment [25] describes the methodology of approaching and estimating the VfM that results by the comparison of the PFI against the conventional procurement options. Indicatively, the quantitative assessment is accomplished with a spreadsheet, by following the instructions of the corresponding guide [26], with the appropriate time risk allocation, which is a critical success factor in PPP projects [27], to those parties that are best able to manage [28]. On the other hand, the CBA methodology as it is presented in the CBA guide for investment projects, issued by the European Commission [29], has six basic steps, including the financial and economic analysis. A remarkable point is that project proposals have to include CBA for all large projects co-financed by the European Community for the 2007-2013 period, where projects considered to be large are those in which the environmental projects budget is over € 25 million and over € 50 million for the other categories [30]. The main purpose of the financial analysis is to use the project cash flow forecasts, in order to calculate the Financial Net Present Value (FNPV) and Financial Rate of Return (FRR). Cash flows arising in different years of the project's lifecycle are calculated by the discount rate (discount flow analysis) in order to adapt to the present value of future flows [31]. Particularly in PPPs, it is recommended that financial analysis should include calculations of the FRR (Kg) and FRR (Kp) indicators, respectively for the public and the private investor. Additionally, economic analysis evaluates the contribution of the project to the economic welfare of the country and is executed on behalf of the whole of society. The key concept is the use of accounting shadow prices, based on the social opportunity cost, instead of observed distorted prices, in order to calculate the Economic NPV (ENPV) and the Economic IRR (ERR). Economic analysis' methodology is divided into five steps: conversion of market to accounting prices, monetisation of non-market impacts, inclusion of additional indirect effects, discounting of the estimated costs and benefits and calculation of the relative indicators, ENPV, ERR and B/C ratio.

3 Economic analysis of wastewater treatment projects

Initially, it has to be mentioned that the most effective way of reducing WT needs and costs, is to reduce the domestic water consumption through an effective water demand management [32]. The second step is the development of WT projects, which have to be planned according to collection and treatment characteristics [33] and the criteria for the selection of the appropriate system are based on factors such as the population density, the produced wastewater volume, the presence of shallow water wells susceptible to wastewater pollution, the soil permeability, the unit cost of wastewater collection and the socio-economic and cultural considerations. However, due to the fact that WT projects should be examined on a case by case basis, the critical point is to distinguish the resulting benefits of the project. Specifically, in dealing with the impact of waste water, boundaries for downstream effects should be clear, either including the area affected immediately, or consider the impacts on irrigation, fishing and drinking water [34]. Moreover, due to the fact that typical environmental impacts are associated with the water quality as well as the soil and groundwater quality, the decrease or increase of the waters' quantity or quality produces some gains or losses in social benefits, while the economic value measurement uses actual related market prices. In the case that market prices do not exist, relevant approaches should be followed. The most common approach is the use of the contingent valuation (CV) method, which is a survey-based method frequently used for placing monetary values on goods and services not bought and sold in the marketplace, for the evaluation of a consumer's willingness to pay for different product/services attributes. Additionally, in the non-market goods appraisal, there is also the Benefit Transfer Approach specifically for environmental goods and services [35], while approaches are also suggested in valuing time, health benefits, landscape or water [36].

4 Case study: evaluation of a wastewater treatment project

The target of this study is to test a hypothetical WT project in both the above evaluation methods, VfM quantitative assessment and CBA, and to combine them suitably by developing a process that could be used in funding evaluation, during a project's feasibility stage. Indicatively, the case study aims to evaluate alternative funding scenarios. The market price, as well as the standard conversion factor and conversion factors that the specific case study uses, are equal to the values of the European CBA guides' relevant case study. Moreover, the initial investment cost is divided equally in the implementation phase's years, while data used have rounded prices for the sake of simplicity. The project is an investment in the field of waste water treatment, for the reuse of well purified waste water for multiple purposes after an intensive tertiary treatment process. It takes place in a member state of the EU and includes the construction of a new water purifier for a city of 200.000 residents in the initial year, while the population grows with an annual rate of 0.5%. Currently, wastewater is discharged untreated into the river crossing the city and part of the water supply is obtained through wells, subjecting the groundwater to an over-abstraction. For this reason, the local aquifer has been depleted, and its hydro geological level has been considerably lowered in recent years. Public authorities decided to examine the funding alternatives of the preferred option, which is the construction of the new water purifier, in order to decide about the funding rates of each participant, i.e. public and private financing in combination with the co-financed EU's grant.

4.1 Assumptions: general

The specific project includes a two year implementation and the 18 year operation of the WT system phases and only financial inflows and outflows are considered.

4.1.1 Initial investment, expenditures and revenues

The cost of the initial investment is estimated to be €30 million, with a 4,5% escalator rate. This is divided into 10% for the feasibility and technical studies, 5% in land expropriation, 25% in labour, 20% in materials for civil works, 10% in transports and rentals and 30% in electromechanical equipment. This cost is 10% higher under the PPP option, as more risk is transferred. The annual operational expenditures (employment) for the initial year of the operational phase are estimated to be € 630.000 in Present Value (PV), which follows a 3,5% escalator rate. On the other hand, the non employment operational expenditures are €200.000/year and the Life Cycle costs, i.e. the annual basis investments during the contract period to maintain the asset so that it remains fit for its intended purpose, are estimated to be € 450.000 for the 1st year, which both follow a 2,5% escalator rate. These costs are met by the private contractor as outflows, while inflows are the annual payments from the public sector. Taking into consideration that the actual daily water supply is estimated to be 190 lt/resident, which has a reduction factor of 0,8 due to water network leakages, and that the purification charge will be €0,32/m^3, expected revenues are: 200.000x190x365x0,8/1000x0,32 = 3.550.720 € / 1st year, altered by 2% annually (1,5% the inflation and 0,5%).

4.1.2 Discount rate

According to the European Guide to CBA, the discount rate that is suggested for investment calculating in the Euro zone during the 2007-2013 period is 5%. Notwithstanding the above rate may vary depending on macroeconomic conditions, or depending on the type of investment, e.g. in PPP projects. For the present project 6% will be taken as the nominal discount rate, based on the Green Book real discount rate of 3,5% [36], and GPD deflator assumption of 2,5%.

4.1.3 Positive and negative externalities

Critical factors that project examiners should take into account are the positive and negative external impacts from the WT plant operation, i.e. the costs and benefits arising to the users, relative costs and benefits for the water resource itself and for the environment in general. Due to the fact that the water supply services is a classic case of a monopoly market, the revenues collected by the owner, even if corrected by means of appropriate conversion factors, do not represent the project's social benefits. Indicatively, some recognised externalities are presented:

4.1.3.1 Negative externalities In the local area there are costs due to the noise, odours, and aesthetic and landscape impacts of the plant. However, for the sake of simplicity, the present case study assumes that the estimated hedonic price is 1.000.000/year, equal to the difference between the market value of the rent for the buildings in the area before the plant is built and the value after the plant is built.

4.1.3.2 Positive externalities The main benefit arising from the specific project's implementation is the groundwater resource saving with the protection of the local hydro-geological level, as well as the generation of many positive environmental effects. Taking an accounting price of € 0,6 per cubic meter of treated water, the benefits are valued at € 6.657.600/year. Additionally, with the use of an accounting price of € 0,7/m^3 for the environmental protection of water and land and the safeguarding of human health and the integrity of human species, environmental benefits have an estimated value of € 7.767.200/year.

4.2 Quantitative VfM assessment

In the case of the quantitative VfM assessment process, which includes the comparison of the Conventional Procurement (CP) against the PFI options, the economic impact, either positive or negative, of the indirect VfM factors should be taken into account. However, only the factors that are likely to arise differentially under one of the tested options are calculated. In the present case study, it is assumed that the PFI option has a 2 million NPV against the CP option (due to risk allocation, design quality, etc). Additionally, the lifecycle cost is taken into account for both options, which represent the cost that is invested during the lifecycle of the project, so the asset remains fit for its intended purpose. The input sheet for the specific case study is illustrated in Table 1. The result of the spreadsheet for a pre tax target Equity IRR of 18%, is the "Indicative" VfM value of 22,70% in favour of the PFI Option, i.e. the PFI is expected to have better VfM than the CP option. Furthermore, the point analysis shows that the switching value of capital expenditure factor is -30%, while the relative switching value of the unitary charge is +32%. These values lie outside the default benchmark tolerances of -5% and 3%, respectively. However, this process does not take into account the revenues that arise by the charging of users, since the payments to the SPV by the state through the unitary charge are adjusted according to the pre tax equity IRR that is used each time. On the other hand, the revenues arising from the investment process, as well as the co-financing rate by other organizations, are considered during the CBA. The process flowchart for funding evaluation with the combination of the above methodologies is illustrated in figure 1.

4.3 Cost benefit analysis

Taking into consideration the revenues that the owner will have by the users' charging, it is proposed that project examiners should evaluate alternative funding rates of the initial investment scenarios in the decision making process. Due to the fact that the project will take place in a EU member state, it is crucial that the initial estimation of the maximum amount to which the co-financing rate of the priority axis applies, with the use of funding gap rate, follows the funding gap method. In the present case, the funding gap rate is calculated to be 11,22%, i.e. the maximum EU contribution is $30x10^6x0,1122 = 3,366x10^6$€.

4.4 Funding evaluation model

The suggested process includes the initial calculation of the investment's indicators under a base case scenario, taking into consideration the maximum EU funding and further the analysis of these indicators under different funding rates in the initial investment.

4.4.1 Base case and alternative scenarios

The EU contributes to the initial investment funding at 11,22% and the remaining 88,78% is covered equally: 44,39% for each of the public sector and the SPV. Indicatively, public funding includes an amount of € 6.658.500 (22,195% of investment costs) by equity, while the remaining 22,195% is provided by a national or regional fund. Additionally, the private financing (€ 13.317.000) is given by equity for 10% of the amount (€ 1.331.700) and by loan for the other 90% (€ 11.985.300), where the loan has a 6% interest rate with an amortization period of 10 years. The service fee paid to SPV is set at € 0.256 per cubic metre of treated water. Calculations of the returns on Local Public capital and Private Equity and the economic analysis of the base case scenario are illustrated in Table 2. For these calculations, it has been taken for granted that the rest of the initial investment, which is not covered by EU funding, is divided equally (44,39-44,39%) between public and private participants. The next step is to analyze various levels of public and private participation, while maintaining full use of the EU funding.

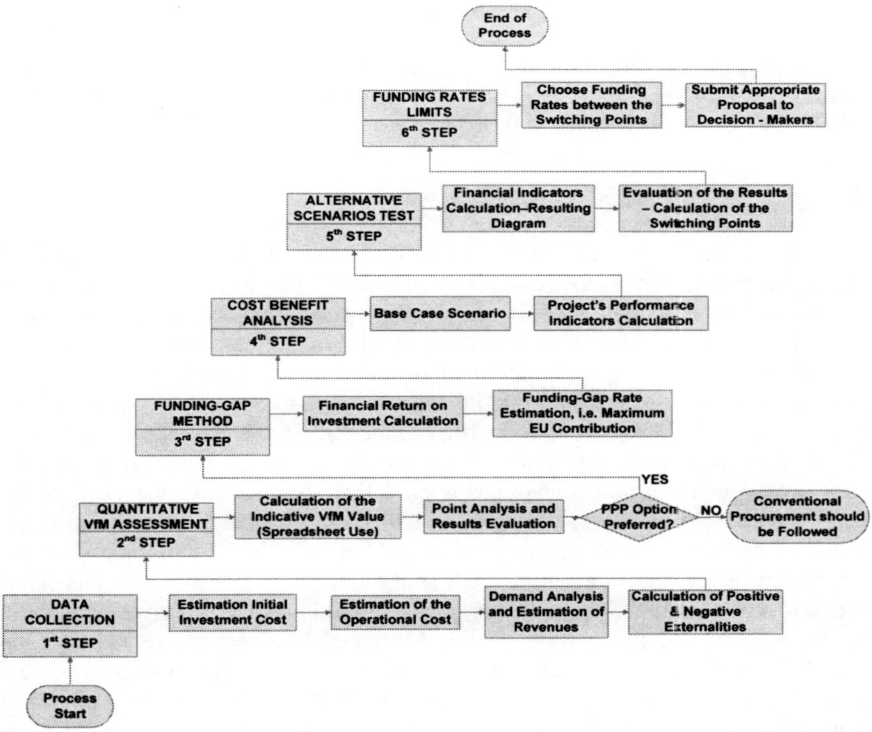

Figure 1: Funding evaluation process flowchart for the implementation of wastewater treatment projects through public private partnership.

4.4.2 Calculations

Calculations of the project's performance indicators under the alternative funding scenarios are also illustrated in Table 2. Furthermore, the resulting diagram, which is presented in Figure 2, demonstrates that the four alternative scenarios result in negative financial indicators either for the public or the private participant. Taking into consideration that the financial indicators of both participants should have positive values, the switching points' calculations are critical, as they reflect the upper and lower limits of the public and private organizations funding rates, as presented in Figure 3. Following the trial and error method, the evaluation process includes the calculation of the switching point's value. For the present case study, as shown in Figure 3, the first switching point value is 31,38% for the SPV and 57,4% for the public sector (27,8% by national or regional fund and 28,7% by local public capital) and the second switching point value is 63,18% for the SPV and 25,6% for the public sector (12,8% by national or regional fund and 12,8% by local public capital). These values present the upper and lower limits of the funding rates that the public and private participants should contribute in the initial investment, so project examiners have to go further in the risk assessment of the CBA, by choosing funding rates between these limits.

4.4.3 Process flowchart

The total process flowchart is illustrated in Figure 1 and presents a new algorithmic type model, which could be used in the funding evaluation process of WT projects implementation through PPPs.

5 Conclusions

The current study examines the initial funding estimation process executed by project examiners during the feasibility stage of WT projects and develops a new algorithmic type model, which follows the option analysis. The new model presents the project examination process in the procurement option under PPPs and specifically the implementation of BOT contracts, where both public and private sectors participate in the initial investment funding. The model is divided into six steps. Initially, it includes the collection of the appropriate data and further the quantitative VfM assessment process in order to demonstrate that the PPP procurement option includes enough VfM. Additionally, the new model takes into consideration that the project will be implemented in a EU member state, so the CBA methodology is used, which includes the funding-gap rate estimation. Later, the financial as well as the economic analysis of a base case scenario are examined, where the rest of the initial investment is covered equally by the public and the private sector. Moreover, the algorithm includes the calculation of various levels of funding rates for both participants and indicatively the switching points, where the FNPV indicators of public and private sectors are positive, i.e. both FRRs are equal to or greater than the discount rate. Switching points resulting from process present the upper and

Table 1: Input sheet, PFI value for money quantitative assessment (adopted by: HM treasury, evaluation spreadsheet, "the spreadsheet").

General

Timings	*(Yrs)*	*Rates - Escalators & Discount*	*Rates (%)*	*Base Year*
Contract period	20	CapEx escalator	4,5%	0
Initial CapEx period	2	OpEx (non employment)	2,5%	0
Year when OpEx is first incurred	3	OpEx (employment) escalator	3,5%	0
Proportion of UC in initial CapEx period payment (%)	50%	Unitary charge escalator	50%	0
		Nominal discount rate	6,09%	NA

Costs *Whole Life*	*CP*	*OB Pre (%)*	*OB Pos (%)*	*PFI*	*OB Pre (%)*
Initial CapEx (€ '000)	27.000	10%	30%	30.000	10%
Lifecycle costs at each LC date (€ '000)	2.700	10%	30%	450	10%
Lifecycle intervals (yrs)	10	NA	NA	1	NA
OpEx (non employment)(p.a.) (€ '000)	200	10%	20%	200	10%
OpEx (employment per person (p.a.)(€ '000)	25,2	NA	NA	20	NA
OpEx (employee number)	20	NA	NA	20	NA
Transaction					
Public sector (€ '000)	500	10%	10%	750	10%
Private sector (€ '000)	0	0%	0%	1.077	10%

Third Party Income	*CP*	*OB Pre (%)*	*OB Pos (%)*	*PFI*	*OB Pre (%)*
Income (p.a.) (€ '000)	0	0%	0%	0	10%

Flexibility	*CP*	*PFI*
Scope change year	10	10
Probability factor (%)	50%	50%
Level of scope change (%)	50%	50%
Premium flexibility factor (%)	0	10%

Pre Tax IRR Targets	
High	18%
Medium	15%
Low	13%

PFI Funding	
Gearing (%)	90%
Sterling swap rate (%)	5,15%
Credit spread (bps)	10
Bank margin (bps)	100
Tail for bank debt (yrs)	2
Commitment fee (bps)	50
Upfront fee (bps)	90

Indirect VfM Factors	*CP*	*PFI*
Amount (NPV)(€ '000)	0	2.00

Tax	*CP*	*PFI*
CP adjustment factor (%)	6%	NA

Lifecycle Related Adjustments	
Lifecycle / residual cost benchmark	50%
CP lifecycle VfM adjustment if lower than benchmark	40%
CP lifecycle VfM adjustment if higher than benchmark	40%
CP residual cost factor if lower than benchmark	70%
CP residual cost factor if higher than benchmark	35%

Table 2: Calculations – funding scenarios.

Scenarios	1st	2nd	Base Case	3rd	4th
EU Contribution	11,22%	11,22%	11,22%	11,22%	11,22%
Public Contribution					
Local Public Capital	12.750.000 € (42,5%)	9.750.000 € (32,5%)	6.658.500 € (22,195%	3.567.000 € (11,89%)	567.000 € (1,89%)
National/ Regional Fund	12.750.000 € (42,5%)	9.750.000 € (32,5%)	6.658.500 € (22,195%	3.567.000 € (11,89%)	567.000 € (1,89%)
Total	**25.500.000 € (85,0%)**	**19.500.000 € (65,0%)**	**13.317.000 € (44,39%)**	**7.134.000 € (23,78%)**	**1.134.000 € (3,78%)**
Private Contribution					
Equity	113.400 € (0,378%)	713.400 € (2,378%)	1.331.700 € (4,43%)	1.950.000 € (6,50%)	2.550.000 € (8,5%)
Loan	1.020.600 € (3,402%)	6.420.600 € (21,402%)	11.985.300 € (39,96%)	17.550.000 € (21,402%)	22.950.000 € (76,5%)
Total	**1.134.000 € (3,78%)**	**7.134.000 € (23,78%)**	**13.317.000 € (44,39%)**	**19.500.000 € (65,0%)**	**25.500.000 € (85,0%)**
FNPV(Kg)	**- 3.795.000**	**-1.045.000**	**+1.789.000**	**+4.623.000**	**+7.373.000**
FRR(Kg)	**+1,722%**	**+4,571%**	**+9,214%**	**+19,050%**	**+88,600%**
FNPV(Kp)	**+15.269.000**	**+10.124.000**	**+4.823.000**	**-470.000**	**-7.793.000**
FRR(Kp)	**+ 390,00%**	**+69,80%**	**+ 17,97%**	**+5,258%**	**-2,437%**
ENPV	**+ 124.988**	**+ 124.988**	**+ 124.988**	**+ 124.988**	**+ 124.988**
ERR	**+ 46,138%**	**+ 46,138%**	**+ 46,138%**	**+ 46,138%**	**+ 46,138%**

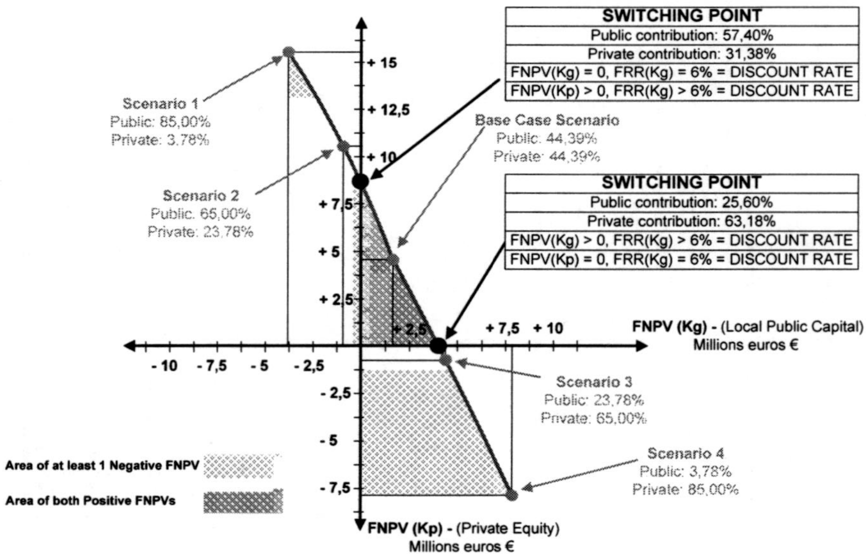

Figure 2: Evaluation of the alternative funding scenarios (switching points) for public and private funding in the initial investment scenarios.

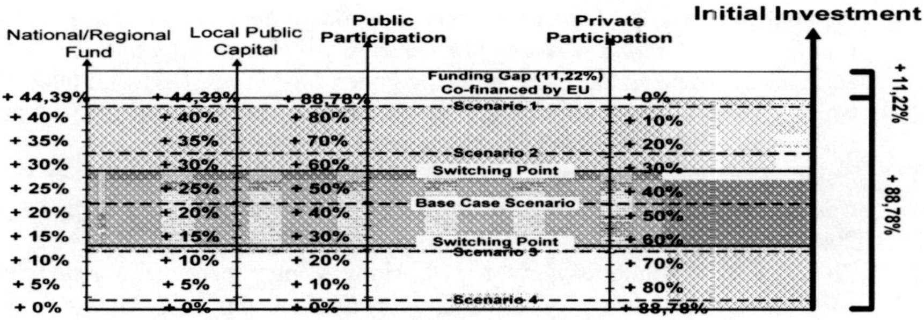

Figure 3: Results of the evaluation process.

lower limits, which include the possible values of the funding rates that the public and the private sectors should contribute in the initial investment. The new model is tested in a hypothetical case study, where the implementation through PPP of a WT project is examined. The model can be used by the project examiners during the feasibility stage of a project, in order to submit the appropriate proposals to decision makers about the funding rates that the participants should contribute in the project's initial investment.

References

[1] Pongsiri, N., Regulation and public private partnerships, *International Journal of Public Sector Management,* **15(6)**, pp. 487-495, 2002.

[2] Jones, R. & Noble, G., Managing the Implementation of Public-Private Partnerships, *Public Money and Management*, **28(2)**, pp.109-114, 2008.

[3] Reeves, E., Public Private Partnerships in the Irish Roads Sector: an Economic Analysis, *Research in Transportation Economics*, 15(1), pp. 107-120, 2005.

[4] Grimsey, D. & Lewis, M.K., Evaluating the risks of public private partnerships for infrastructure projects, *International Journal of Project Management*, **20(2)**, pp. 107-118, 2002.

[5] Bovaird, T., Public–private partnerships: from contested concepts to prevalent practice, *International Review of Administrative Sciences*, **70(2)**, pp. 199-215, 2004.

[6] Evans, G. & Bowman D., *The Challenge of Public Private Partnerships: Learning from International experience*, Edward Elgar publ, pp.64-65, 2005.

[7] World Bank Group, PPP in Developing Countries, 2007 data results from the PPI Project Database, The World Bank Group, Public Private Infrastructure Advisory Facility, pp. 13-19, http://ppi.worldbank.org/

[8] Federal Highway Administration, *Synthesis of public private partnership projects for roads, bridges and tunnels from around the world 1985–2004*, U.S. Department of Transportation, Washington, D.C., pp. 9-10, 2005.

[9] Panggabean, A.T.P., *Expanding Access to Basic Services in Asia and the Pacific Region: Public-Private Partnerships for Poverty Reduction*, Asian and Development Bank, ERD Working Paper, No 87, pp. 1, 7, 14, 2006.
[10] Chiu, Wai-ip, *How to Launch a Successful BOT Project*, Edmond University of Hong – Kong, Thesis submitted, p. 34, 1998.
[11] Asian Development Bank, *Technical Assistance for Public Private Community Partnerships in Urban Services for the poor*, A. 2, pp. 1-2, 2000.
[12] Zhang, X-q. & Kumaraswamy, M. M., Hong Kong experience in managing BOT projects, *Journal of Construction Engineering and Management*, **127(2)**, pp. 154–162, 2001.
[13] Haarmayer, D. & Mody, A., *Financing Water and Sanitation Projects – The Unique Risks*, The World Bank Group, Finance, Private Sector and Infrastructure Network, 151, pp. 1-4, 1998.
[14] European Commission, *Guidelines for Successful Public – Private Partnerships*, Directorate – General Policy, p.14, 2003.
[15] European Commission, *Green Book on Public Private Partnerships,* Directorate – General Policy, COM 327, final, 2004.
[16] United Nations, *Guidebook on Promoting Good Governance in Public Private Partnerships*, Economic Commission for Europe, pp. 68-69, 2008.
[17] HM Treasury, *How to appoint and work with a preferred bidder*, Technical Note No. 4, Treasury Taskforce, Private Finance, HMSO, London, 1999.
[18] Office of Government Commerce, *Green public private partnerships*, Norwich, pp.4-5, 2002.
[19] Grimsey, D. & Lewis M.K., Are Public Private Partnerships value for money? Evaluating alternative approaches and comparing academic and practitioner views, *Accounting Forum*, **29(4)**, 2005, pp. 345–378, 2005.
[20] HM Treasury, *Value for money assessment guidance*, London, 2004.
[21] Spackman, M., Public–private partnerships: lessons from the British approach, *Economic Systems*, 26**(3)**, pp. 283–301, 2002.
[22] HM Treasury, http://www.hm-treasury.gov.uk/d/pfi_signed_projects_ list.xls
[23] HM Treasury, *PPP's, the Government's Approach*, p.10, London, 2000.
[24] Department for Transport, *Green Public Private Partnership*, UK, p.4, 2002.
[25] HM Treasury, *Value for Money Assessment Guidance*, London, 2006.
[26] HM Treasury, PFI Value for Money Quantitative Assessment, Evaluation Spreadsheet, http://www.hm-treasury.gov.uk/d/vfm_qe_spreadsheet 0307.xls
[27] Zhang, X., Critical Success Factors for PPPs in Infrastructure Development, *Journal of Construction Engineering &Management*, **131(1)**, pp. 3-14, 2005.
[28] Van Ham, H. & Koppenjan, J., Building PPPs: Assessing and managing risks in port development, *Public Management Review*, **4(1)**, pp. 593–616, 2002.

[29] European Commission, *Guide to Cost–Benefit Analysis of Investment Projects*, pp. 13 – 15, 2008.
[30] European Union Official Journal, *Council Regulation (EC) No 1083/2006*, of 11 July, 2006: Chapter II, Section 2, Article 39-40, 2006.
[31] Project Management Institute, *PMBOK*, 4th ed, p.168, 2008.
[32] United Nations, *Guidelines on Municipal Wastewater Management*, UN Environmental Programme, pp.48-51, 2004.
[33] Organisation for Economic Cooperation & Development, *Handbook for Appraisal of Environmental Projects Financed from Public Funds*, Environmental Finance, pp. 64 – 66, 2007.
[34] Asian Development Bank, *Guidelines for the Economic Analysis of a Project*, Economics and Development Resource Centre, p. 167, 1997.
[35] Pearce, D., Atkinson, G. & Mourato, S., *Cost-Benefit Analysis and the Environment: Recent Developments*, OECD publishing, pp. 253-267, 2006.
[36] HM Treasury, The Green Book: Appraisal and Evaluation in Central Government, London, *Annex 2: Valuing non market impact*, pp. 57-67, 2004.

Section 4
Natural resources management

Payments for environmental services (PES): contribution to Indigenous livelihoods

R. Greiner
Charles Darwin University, Australia

Abstract

Indigenous people living in Australia's tropical savanna landscapes are increasingly looking towards income opportunities from the provision of environmental services as an avenue for economic development and improvement in socio-economic conditions. A number of programs, which can be classified as payments for environmental services, support the existence and operation of Indigenous land and sea management groups, also referred to as Indigenous ranger groups. Rangers undertake a portfolio of activities, including feral animal and weed control, biodiversity monitoring and protection, fire management, and biosecurity and border protection assignments. This paper reviews the extent to which current activities contribute to Indigenous livelihoods and discusses their likely growth potential and livelihood contribution in the future.

The terms 'Indigenous' and 'Traditional Owners' are capitalised in this paper as they refer to descendants of the original inhabitants of Australia.
Keywords: payments for environmental services, livelihoods, Indigenous people, assessment, Australia.

1 Introduction

Australia's tropical savannas encompass approximately one quarter of the continent (around 1.9 million km^2) and span northern Queensland (Qld), the Northern Territory (NT) and northern Western Australia (WA). The population density of the tropical savannas is low (<0.2 persons per km^2), with fewer than 500,000 persons living in the region [1,2]. A large proportion of the resident population is Indigenous. Indigenous people are those who self-identify as being of Aboriginal or Torres Strait Islander descent and are accepted by the

www.witpress.com, ISSN 1743-3541 (on-line)
doi:10.2495/EEIA100141

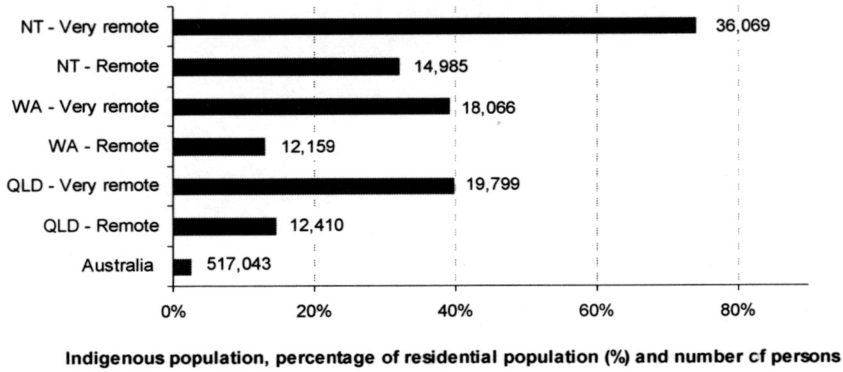

Figure 1: Indigenous population in 'remote' and 'very remote' regions (compiled from [3]).

community as such. Indigenous people account for up to 75% of the residential population in some of the 'very remote' areas of the tropical savannas [3], fig 1.

'Remote' and 'very remote' areas of the NT and Qld are amongst the most socio-economically disadvantaged areas of Australia [5]. Indigenous populations in particular tend to be characterised by low workforce participation and high unemployment [6]. The employment profile of the 27,950 Indigenous persons who were employed across northern Australia at the date of National Census of Population and Housing in 2006 [4] shows high levels of employment particularly in public service and health-related industries, fig.2. The third largest category is 'unspecified', which may relate to people participating in the Community Development Employment Projects (CDEP) scheme, which in 2006 was still generally operational. The CDEP scheme enabled members of Indigenous communities, particularly in remote and very remote areas, to exchange unemployment benefits for opportunities to undertake paid work and training in activities managed by a local Indigenous community organisation. The 'unspecified' category could possibly also relate to employment in the natural resource management sector, which does not fit well with the industry profile adopted for census data collection.

A major factor is the virtual absence of the commercial sector, other than the mines operating in some places. Being employed in natural resource management, working on country and being paid for providing environmental services is thus an appealing concept. It deserves major consideration in the discussion about future development trajectories for tropical savannas [2], particularly since these landscapes produce among the highest per-hectare ecosystem values in Australia [7]. The abundance of natural capital—in terms of size and quality—provides market opportunities but also poses a number of challenges [8].

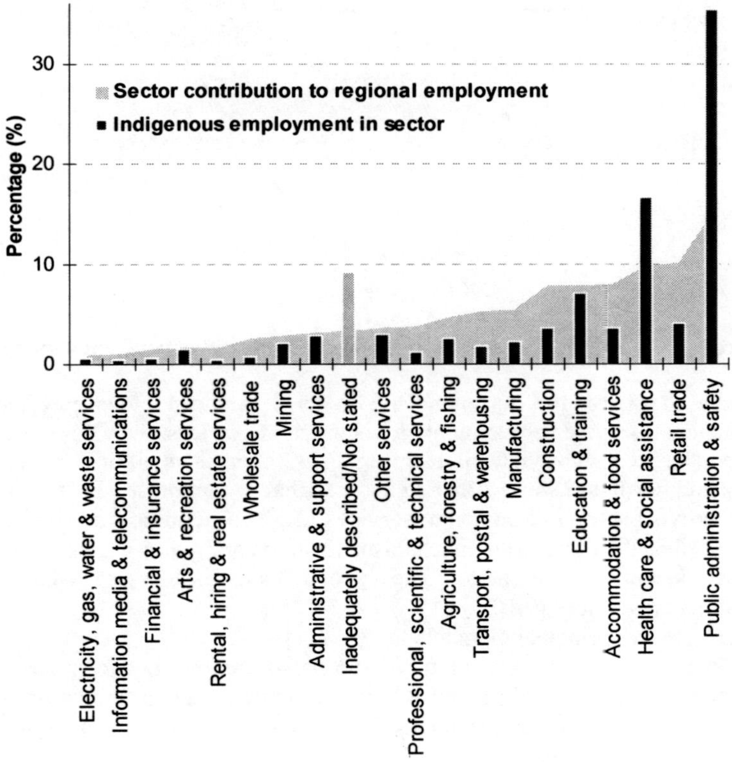

Figure 2: Employment in northern Australia, by industry (compiled from [4] for Kimberley, NT, North Qld and Far North Qld).

Following some necessary clarification of terminology in section 2, the paper goes on to describe Indigenous land and sea management initiatives and illustrates existing PES schemes of relevance (section 3). It concludes with an integrated assessment of the likely livelihood benefits from such schemes for Indigenous Australians in northern Australia (section 4).

2 Payments for environmental services

Environmental services are conceptually distinct from ecosystem services. The notion of "ecosystem services" is outcome-based and focused on the wellbeing benefits provided to society from natural capital. In contrast, the notion of "environmental services" is input-based and focuses on the efforts undertaken by actors to generate environmental improvements and improved natural capital [9], fig.3.

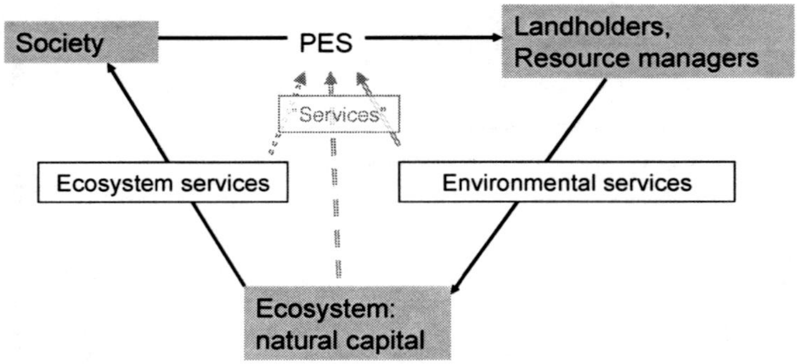

Figure 3: PES in the context of environmental and ecosystem services [8].

In the literature, the acronym PES is used to mean either payments for environmental services or payments for ecosystem services, with little distinction between the two. A review of a dozen PES schemes operating in Australia, Europe and north and south America showed that in fact most PES schemes pay for the provision of environmental services [8]. The definition adopted here is, therefore, that PES is a price mechanism for translating non-market values into financial incentives for actors (i.e. people and businesses) who provide environmental services [9,10].

There are five principal elements to PES [11]:

1. PES are based on a voluntary, negotiated contract between suppliers of environmental services and a buyer. Suppliers have a choice of entering into contractual arrangements but if they do they must provide the environmental service.
2. The service must be well-defined. If it is not directly measurable—as is the case with most environmental outcomes—surrogate measures are required that are directly linked to the provision of the service.
3. There needs to be at least one buyer. The buyer can be a government, acting on behalf of a community, or a private entity or non-government organisation. There are often brokers or intermediaries involved in functioning environmental or ecosystem services markets.
4. There also needs to be at least one supplier/provider. In most cases, there will be many providers.
5. Payments are made conditional on the supplier meeting the service delivery specifications to ensure the effectiveness of the scheme.

Supply of environmental services is principally driven by the price received for environmental services, opportunity costs, risk and the institutional framework, fig.4. If, for example, statutory requirements demand a high level of environmental service provision as part of an actor's environmental duty of care, the scope for PES is reduced. Demand is principally driven by the utility that users derive from ecosystem services, the number of users, and their willingness to pay. Whether a market will emerge is further influenced by the clarity of trading rules and the level of transaction costs, which can be significantly

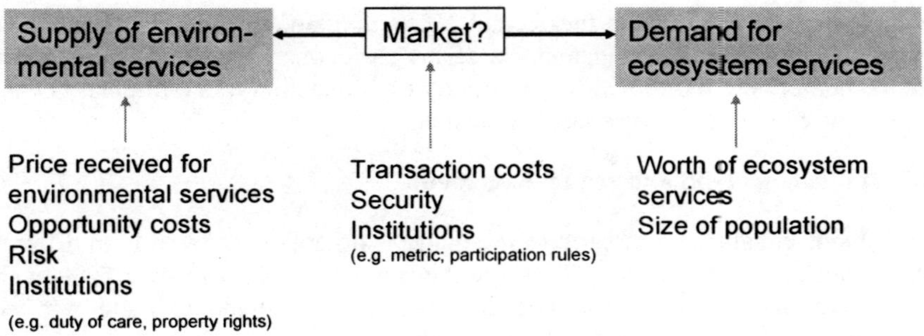

Figure 4: Determinants of markets for environmental services [8].

reduced in monopsony (single buyer) situations [12]. Because of the inherent complexity and characteristics of the natural and social systems that they interact with, PES might not always be the most effective policy tool for the provision of ecosystem services [12].

3 PES involving Indigenous people in northern Australia

3.1 Context

In 2000, the Council of Australian Governments (COAG—i.e. the Australian Government and the seven state and territory governments) agreed to endorse a National Action Plan to address the natural resource issues of salinity and water quality [13]. In doing so, it effectively instituted a nation-wide PES scheme with the intention to enable and motivate regional communities to use coordinated and targeted action to address a number of environmental degradation problems through landscape-scale change. Regional natural resource management (NRM) groups were set up across the country to act as coordinators and brokers between the government—as monopsony buyer—and landholders as the providers of environmental services. Importantly, following the recognition of Native Title in 1993, Traditional Owners were recognised as key stakeholders in NRM and a formal avenue established for their engagement in the governance, management and on-ground dimensions of regional NRM. The stated objective of this engagement is to capture the significant economic and socio-cultural benefits associated with NRM, specifically in areas where there are few commercial avenues for improving socioeconomic well-being [14].

Indigenous people refer to NRM as 'caring for country'. Caring for country is an inclusive concept in that 'country' includes both land and sea country and 'caring' relates to country as well as people. Over the past decade, Indigenous engagement in NRM has grown through increasing involvement of Traditional Owners (i) organisationally, through the establishment of Indigenous land management facilitator positions within NRM groups and the development of Indigenous land and sea management/Indigenous ranger groups, and (ii) operationally, through an increasing suite of on-ground programs, which now operate within and outside the formal NRM domain. In northern Australia in particular, Indigenous engagement in caring for country has been strong due to the comparatively high levels of continuity of connection of Traditional Owners with their country since European settlement.

3.2 Indigenous land and sea management

Indigenous engagement in caring for country is organised differently in different states and territories. In the NT, where Aboriginal people own over 50% of the land mass and 84% of the coastline, some 46 Indigenous land and sea management groups with around 500 members were operating in 2007 [15]. The groups are principally supported in their operations by the Indigenous land councils while the 2003-founded North Australian Indigenous Land & Sea Management Alliance (NAILSMA [16]), which also operates in northern Qld, and northern WA, plays an important advocacy role. In Qld, Indigenous caring for country activities tend to align with regional NRM groups, which—under the auspices of the Statewide Indigenous Network Coordination Project—employ Indigenous facilitators and fund traditional owner involvement in caring for country activities [17]. Some traditional owner groups are self-organised as Indigenous land and sea management or ranger groups, while others operate under state government auspices (e.g. Wild river rangers [18]).

The portfolio of caring for country activities include, *e.g.*, feral animal and weed control, biodiversity monitoring, threatened species protection, fire management and coastal rehabilitation. Some groups conduct surveillance and quarantine related activities. An important aspect of Indigenous land and sea management is the continued growth of social and human capital through training and education programs, policy and planning, advocacy, and community development projects.

Indigenous land and sea management is principally funded by government. Two types of funding exist, what one might call 'core funding' and 'project-based funding'. Core funding comes, *inter alia*, from wage support (such as CDEP), Indigenous infrastructure and activity funding, Indigenous environmental and cultural heritage programs, resources for training and resources for business and industry development [15]. While accountability criteria apply to core funding, it is project-based (or grant) funding that can be interpreted as PES. The following examples serve to illustrate PES-style projects for Indigenous caring for country activities that operate at present in the tropical savannas.

3.3 Removal of marine debris from remote beaches

Circular currents in the waters off northern Australia mean that marine debris (i.e. any manufactured or solid processed material that enters the marine environment from any source [19]) which enters the Gulf of Carpentaria, remains trapped in those currents and accumulates. Even after being dumped on beaches, marine debris often gets re-suspended during high-wind events [20]. Discarded fishing nets, so-called 'ghost nets', pose a key concern as they 'continue to fish' and kill marine wildlife while they remain suspended in the sea [21].

The Ghost Nets Programme commenced in 2004 with the objective to remove discarded fishing gear and other marine debris from coastline around the Gulf of Carpentaria to support biodiversity conservation, and to release trapped animals.

The program is coordinated through the Northern Gulf Resource Management Group, a regional NRM group. Coastal clean-up services are provided by an alliance of Indigenous communities surrounding the Gulf of Carpentaria who refer to themselves 'saltwater people'. Work is carried out by members of the Indigenous rangers groups of participating communities—currently 18 ranger groups across Qld and the NT. In the first 4½ years of the program's existence, Indigenous rangers removed approximately 90 km of ghost nets [22]. The NRM group submitted a bid to the Australian Government's 2009 Caring for Our Country program and secured funding of approximately AU$ 1 million per year for an additional three years for the program.

The Ghostnet Programme illustrates a number of issues associated with government-related funding for PES schemes: (i) Continuity of grants-based schemes is uncertain and difficult to achieve given election cycles, changes in government priorities and competitive allocation of funds. (ii) Funding is generally at a level that does not provide for employment of Indigenous rangers for a singular environmental service: In comparative terms, funding for the Ghost Nets Programme is substantial. However, even if administrative and management costs are small, participating land and sea management groups receive on average less than AU$ 50,000 per annum to cover wages and operating expenses.

3.4 Controlled savanna burning to reduce the level of fire-related greenhouse gas emissions

Wild fires play an important role in tropical savannas. Fire has a major influence in shaping the vegetation, with frequent dry season fires limiting tree recruitment and keeping the canopy open [23]. However, fires consume vegetation and, in doing so, produce CO_2 and other greenhouse gases. While fires are a natural phenomenon, the timing and intensity of fire, and consequently its various ecological impacts, can be significantly influenced by fire management activities. In savanna landscapes, fire management based on traditional Indigenous knowledge and fire management practices, and supported by modern technology, can substantially reduce greenhouse gas emissions and generate biodiversity benefits in comparison to an 'unmanaged' situation, such as that which arose post European settlement [24].

The West Arnhem Land Fire Abatement (WALFA) Project harnesses these relationships [24,25]. The project is a contractual 17-year partnership between the Aboriginal owners and five affiliated Indigenous ranger groups, Darwin Liquefied Natural Gas (LNG), the Northern Territory Government and the Northern Land Council as broker. Through this partnership, Indigenous ranger groups are implementing strategic fire management across 28,000 km^2 of

Western Arnhem Land to offset greenhouse gas emissions from the liquefied natural gas plant in the city of Darwin. Darwin LNG pays the Indigenous land and sea management groups around AU$ 1 million per year to provide this specialised fire management service with an annual sequestration target of 100,000t CO_2-equivalent (CO_2e) relative to a 10-year (1995-2004) baseline of the region's fire-derived greenhouse gas emissions [26]. Between 2005 and 2008, the project has achieved a measurable reduction in destructive wild fires and an estimated average annual reduction in release of greenhouse gases from the contracted area of 122,000t CO_2e per annum [24].

The outcomes achieved by the West Arnhem Fire project, both procedural and operational, have potential application across fire-prone tropical Australia as well as fire-prone tropical regions elsewhere [25].

The WALFA project demonstrates the potential scope for commercially funded large-scale PES schemes. Long-term contracts between companies and services providers can offer high levels of security and continuity for both parties. However, the existence of such schemes depends on government providing policy settings which underpin markets for environmental goods such as CO_2e. Legislative frameworks that set pollution caps and accounting and trading rules fundamentally determine the existence of a market, potential participants, transaction costs and per-unit prices. At an approximate level of payment of $10/tonne CO_2e or approximately $35-46/km^2, even commercial PES schemes of a WALFA nature are unlikely to sustain land and sea management groups in their own right either financially or operationally, as burning activities are seasonally limited.

3.5 Provision of surveillance and quarantine services

Australia's northern coastline is vast and sparsely populated, making it vulnerable to undetected foreign vessels that bypass the usual quarantine checks at Australian borders. There is also a risk that pests could enter Northern Australia from neighbouring countries, using migrating birds, human activities or wind currents as carriers [27].

The Northern Australia Quarantine Strategy was established by the Australian Quarantine and Inspection Service (AQIS) in 1989 and conducts surveys along Australia's northern coastline and in neighbouring countries for early signs of new pests or disease. AQIS engages Indigenous land and sea management groups (20 groups in the NT, 12 in WA, 8-10 in Qld) on a fee-for-service basis to support surveillance and quarantine efforts. Indigenous rangers harvest feral pigs, inspect them for the presence of serious exotic diseases like foot and mouth disease, and take blood samples for further analysis [28].

Such fee-for-service arrangements offer continuity of funding since the need for services is ongoing and there are no alternative service providers. Given skills required to undertake tasks, this may provide 'permanent' jobs for some members of Indigenous land and sea management units and offer a stable contribution to the funding of ranger groups.

4 Assessment of PES: contribution to Indigenous livelihoods

PES schemes typically operate where natural assets are scarce and/or under immediate threat. The need for relative scarcity poses a challenge for the savannas of northern Australia, where natural capital is relatively abundant and population is sparse: those who provide environmental services need to largely cater to the demand for ecosystem services from people elsewhere. Not everybody who provides an environmental service can expect payment "since services that are neither highly valuable and/or not threatened are unlikely to find buyers" [11, p.5]. However, in the abundance also lies a market opportunity in that buyers aiming to protect existing but locally and/or globally threatened services may want to anticipate emerging threats and future cost rises—similar to the concept of ecological insurance against uncertain futures. If they delay purchasing services until after changes have occurred, stocks of natural capital will be low and ecosystem services more scarce (some may already have been irreversibly lost) and the cost of delivering environmental services will be more expensive [8]. Also, in the case of greenhouse gases, abatement costs in savanna landscapes may be lower than elsewhere.

In the tropical savannas context, matters of size, low opportunity costs from forgone production, the pristine state of many areas, and uniqueness, enhance the marketability of existing ecosystem services. There is an Indigenous labour force scattered throughout the landscape in settlements and outstations, where employment and commercial opportunities are limited. Over the past decade, Indigenous land and sea management has successfully harnessed this combination and grown into a successful avenue for employment and economic development that builds on Indigenous people's skills, interests, location and land ownership.

Comparatively speaking, Indigenous NRM is not a large industry, with approximately 500 workers employed in the NT. At an individual level, however, it can be argued that every one of these jobs makes a difference to an Indigenous person and, possibly, a family. In addition to employment and income, these jobs generate pride, recognition, skills, work ethic and other social values which may be able to help address some causes of the socio-economic disadvantage faced by 'remote' and 'very remote' Indigenous communities. It may provide a foundation for other initiatives and more employment in other industries. As the population of Indigenous communities, particularly in remote areas, grows rapidly [2], the relative contribution that caring for country can provide to community livelihoods is likely to diminish over time.

There are limits to the growth of this new industry. Caring for country is essentially a public sector industry with government—albeit through various agencies and possibly through different levels of government—as the single or dominant purchaser of services. As such both 'core' funding and grant funding are subject to government persuasion, election cycles, and changing priorities and policy manifestations. While there are multiple sources of grant funding, transaction costs for ranger groups associated with competitive funds are high and the ability of groups to demonstrate their capacity to deliver is fundamental for their viability. Ongoing investment in ranger education and skills development are critical, yet this cannot be funded by PES schemes. Consequently, groups will continue to rely on 'core' government funding to be operational and competitive.

The WALFA project has shown that commercial PES arrangements for environmental services provided on Indigenous lands by Indigenous rangers are feasible. However, the opportunity for expansion of such arrangements—in particular as they relate to greenhouse gas abatement—is uncertain. The WALFA project can be considered a philanthropic project, since there are currently no legal requirements for greenhouse gas emitters in Australia to off-set their emissions. Large-scale fire management across tropical savannas under commercial arrangements will only eventuate if there is a market. A market only operates if government policy has set the necessary institutional foundations. If the Australian Government succeeds in introducing a greenhouse gases cap-and-trade system, the future for such schemes across the tropical savannas is bright.

Acknowledgements

I would like to thank Heather Aslin and Julie Crough for helpful comments on an earlier draft of this paper.

References

[1] Woinarski, J., Mackey, B., Nix, H., and Traill, B. 2007. The nature of northern Australia - Natural values, ecological processes and future prospects. ANU Press: Canberra

[2] Garnett, S., Woinarski, J., Gerritsen, R., and Duff, G. 2008. Future options for North Australia. Charles Darwin University Press: Darwin.

[3] Australian Bureau of Statistics. 2008. Experimental Estimates of Aboriginal and Torres Strait Islander Australians, Jun 2006. Cat. No. 3238055001DO004_200606. Commonwealth of Australia: Canberra.

[4] Australian Bureau of Statistics. 2007. Industry of Employment by Indigenous Status. Cat. No. 2068.0 - 2006 Census Tables. Commonwealth of Australia: Canberra.

[5] Australian Bureau of Statistics. 2008. Socio-Economic Indexes for Areas (SEIFA). Cat. No. 2033.0.55.001. Commonwealth of Australia: Canberra.

[6] Australian Bureau of Statistics. 2008. Australian Social Trends 2008. Cat. No. 4102.0. Commonwealth of Australia: Canberra.

[7] Costanza, R., d'Arge, R., de Groot, R., Farber, S., Grasso, M., Hannon, B., Limburg, K., Naeem, S., O'Neill, R.V., Paruelo, J., Raskin, R.G., Sutton, P. and van den Belt, M. 1997. The value of the world's ecosystem services and natural capital. *Nature* 387, 253–260.

[8] Greiner, R., Gordon, I. and Cocklin, C. 2009. Ecosystem services from tropical savannas: economic opportunities through payments for environmental services. *The Rangeland Journal* 31, 51–59.

[9] Fisher, B., Turner, R.K., and Morling, P. 2008. Defining and classifying ecosystem services for decision making. *Ecological Economics* 68, 643-653.

[10] Engel, S., Paiola, S., and Wunder, S. 2008. Designing payments for environmental services in theory and practice: An overview of the issues. *Ecological Economics* 65, 663–674.

[11] Wunder, S. 2005. Payments for environmental services: some nuts and bolts. Occasional Paper No 42. Center for International Forestry Research: Jakarta, Indonesia.

[12] Kemkes, R.J., Farley, J., and Koliba, C.J. 2009. Determining when payments are an effective policy approach to ecosystem service provision. *Ecological Economics* doi:10.1016/j.ecolecon.2009.11.032

[13] Keogh, K., Chant, D., and Frazer, B. 2006. Review of arrangements for regional delivery of natural resource management programmes. Ministerial Reference Group for Future NRM Programme Delivery: Canberra.

[14] Altman, J., and Whitehead, P. 2003. Caring for country and sustainable Indigenous development: Opportunities, constraints and innovation. CAEPR Working Paper No. 20/2003. Centre for Aboriginal Economic Policy Research, Australian National University: Canberra.

[15] Putnis, A, Josif, P., and Woodward, E. 2007. Healthy Country, Healthy People: Supporting Indigenous Engagement in the Sustainable Management of Northern Territory Land and Seas: A Strategic Framework. CSIRO: Darwin.

[16] NAILSMA 2009. About NAILSMA. Online at: http://www.nailsma.org.au/about_nailsma/index.html. Accessed: 01/01/2010.

[17] Qld Regional Group Collective. 2009. Indigenous land and sea management in Queensland. Online at: http://www.rgc.org.au/. Accessed: 03/01/2010.

[18] Qld government. 2009. Wild river rangers. Online at: http://www.regionalnrm.qld.gov.au/policies_plans_legislation/policies_strategies/wild_river_rangers.html. Accessed: 03/01/2010.

[19] Coe, J and Roger D. (Eds.). 1997. Marine Debris Sources, Impacts and Solutions. Springer-Verlag, New York

[20] Carpentaria Ghost Nets Programme. 2009. About the program. Online at: http://www.Ghost Nets.com.au/about.html. Accessed: 01.01.2010.

[21] Anon. 2006. Marine debris in Northern Australia: A short history of work on the problem. Online at: http://www.Ghost Nets.com.au/data/History_marine_debris_northernAustralia.pdf. Accessed: 01/01/2010.

[22] Carpentaria Ghost Nets Programme. 2009. Survey results. Online at: http://www.Ghost Nets.com.au/data. Accessed: 01.01.2010.

[23] Dyer, R., Jacklyn, P., Partridge, I., Russell-Smith, J., and Williams, R. 2001. Savanna Burning: understanding and using fire in northern Australia. Tropical Savannas CRC: Darwin.

[24] Russell-Smith, J., Whitehead, P. and Cooke, P. (Eds.) 2009. Culture, Ecology and Fire Management in North Australian Savannas: Rekindling the Wurrk Tradition. CSIRO Publishing: Collingwood.

[25] TS-CRC. 2009. The West Arnhem Land Fire Abatement Project (WALFA). Online at: http://www.savanna.org.au/all/walfa.html Accessed 15/12/09.
[26] Whitehead, P., Purdon, Pl, Cooke, P., Russell-Smith, J., and Sutton, S. 2009. The West Arnhem Land Fire Abatement (WALFA) project: the institutional environment and its implications. P.298-312, in, Russell-Smith, J., Whitehead, P. and Cooke, P. (Eds.) 2009. Culture, Ecology and Fire Management in North Australian Savannas: Rekindling the Wurrk Tradition. CSIRO Publishing: Collingwood.
[27] AQIS. 2009. Northern Australia Quarantine Strategy (NAQS). Online at: http://www.daff.gov.au/aqis/quarantine/naqs. Australian Government Department of Agriculture, Fisheries and Forestry: Canberra. Accessed: 01/01/2010.
[28] Tropical Savannas CRC. 2009. Biosecurity: keeping out unwelcome visitors. Online at: http://www.savanna.org.au/all/biosecurity.html. Accessed: 03/01/2010.

Enhancing natural resource management through payment for ecosystem services

S. Vemuri[1] & J. Gorman[2]
[1]*School of Law and Business, Faculty of Law, Business and Arts (LBA), Charles Darwin University, Australia*
[2]*School for Environmental Research, Institute of Advanced Studies, Charles Darwin University, Australia*

Abstract

There is evidence to suggest that for >60,000 years prior to European colonisation, Indigenous Australians used ecosystem services and managed landscapes sustainably. Since that time and in the wet/dry tropics of the Northern Territory, fire, weeds and feral animals have had, and continue to have, the greatest influence on landscapes and the ecosystem services they provide. Internationally, the links between Indigenous people and their experiences with natural and cultural resource management has been recognized and now have become an important and popular strategy for promoting sustainable development. Indigenous Natural and Cultural Resource Management (INCRM) is particularly relevant in the Northern Territory of Australia and has contributed to the formation of an Indigenous ranger program to manage threats to the landscape in conjunction with customary land management practice from Aboriginal people still living on their country. A closer examination of the success of INCRM suggests that the real challenge is one of managing knowledge pluralism rather than finding a common base for the different sources of knowledge. More recently these landscapes have been subject to a number of additional threats (spread of disease from feral animals, illegal international fishing vessels, exotic ant control, fire abatement etc), which still require management at a local level. Many of these are currently being managed under a payment for environmental service (PES) type arrangement. While PES has great public cost-benefit, it needs to be determined if the model provides the best mechanism for the progress of Indigenous people. This paper is about the intricacies of issues related with linking knowledge of Indigenous people with

WIT Transactions on Ecology and the Environment, Vol 131, © 2010 WIT Press
www.witpress.com, ISSN 1743-3541 (on-line)
doi:10.2495/EEIA100151

provision of ecosystem services and will conclude the ongoing commitment of PES to sustainable development.

Keywords: PES, aboriginal, holistic, governance, aboriginal, rangers.

1 Introduction

This paper examines the current importance placed on Payment for Ecosystem Services (PES) as a tool for enhancement of nature's resources. In so doing, there is a need to take heed from Hicks [1] and note that "if one says that the economist is concerned with the present, that is just another way of saying he is concerned with the past and with the future." In particular the focus of the paper is on the phylogenic conception of evolution of four units of selection based on the taxonomy suggested by Hodgson [2], namely individuals, routines, institutions and systems. The paper suggests that the contemporary importance of PES in Natural Resource Management is a cumulative effect of the process of adaptation of means and ends between agents to achieve pluralistic outcomes. This is partly the reason for acceptance of PES by all participants in the NRM process. Unfortunately such an acceptance does not automatically imply enhancement of NRM.

The Northern Territory of Australia is selected as the study site for examining the enhancement of resource management through PES, as it provides an opportunity to study the integrated nature of the functions of the natural environment as suggested by de Groot [3], with the public policy initiatives such as Caring for Country and Indigenous Ranger programs adopted by the Australian Government through National Heritage Trusts and the National Action Plan for Salinity and Water Quality.

The paper is organized as follows. The first section of the paper describes the challenges facing NRM in the context of Northern Australia. The second section of the paper traces the historical context of PES in the Northern Territory of Australia. The third section of the paper examines the current status of PES related activity. In the final section the paper discusses the future importance of PES in enhancing NRM.

2 Northern territory and NRM

The Northern Territory of Australia is a region with unique natural and social values. There is a great diversity of plants and animals associated with habitats which range from rainforests, mangrove forests, swamps and wetlands in the wet/dry tropics in the north through to hummock grasslands and mulga shrub lands of the desert areas in the south. Culturally there is also great diversity with over 200 Aboriginal communities living in this area with intact spiritual connections and obligations to care for and manage country. Aboriginal people consider country as being 'a living entity with a yesterday, today and tomorrow, with a consciousness, and a will towards life' (Rose [4]). Without people living on the land in remote Northern Territory landscapes there are threats from fire, disease, feral animals, illegal fishing, and bio-security and left unmanaged the very values that make this area unique and such an asset to Australia would be seriously undermined.

Indigenous owned land in the NT includes some of the most bio-diverse lands in Australia (Altman et al. [5]; DNRETAS [6]). These include many of the most intact and nationally important wetlands, riparian zone, forests and rivers and water ways. Despite their relative intactness these lands are at risk of species contraction and face major threats from feral animals, exotic weeds, changed fire regimes, pollution and over grazing (Garnett and Woinarski [7]).

In the Northern Territory (NT) of Australia Aboriginal people are progressively seeking greater empowerment and self-determination of their lands, institutions and affairs. One way to achieve self-determination is through participating in business related activities. Concerted attempts are underway to promote enterprise development. Most economic opportunities exist for Indigenous people through managing their lands and utilizing their plant and animal resources (SRRATRC [8]). Many contributions to the field of sustainability have not only recognized but also acknowledged the significant role of traditional Indigenous knowledge to commercialization of land management activities. Discussions are normally couched along the lines of building capacity to deal with, and integrate, the scientific know-how with traditional environmental knowledge.

Over the years there have been several attempts to address the economic and environmental, challenges in a comprehensive manner. An understanding of these attempts will provide a historical context why PES is considered to enhance NRM.

3 Historical context

Historical events have shaped the way public policy has addressed issues related to NRM. Although there have been variations to the NRM policies there has been an underlying drive that jealously propagates "Shamrocks" in the form of ecology ,economy and society (Handy [9]) in the environment on the one hand and address the "hybrid" nature of the economy (Altman [10]) on the other hand.

The combination of producing Shamrocks in a hybrid context has created in essence a "*Leitbild*" (Pearson and Gorman [11]). This is most evident in the way NRM strategies have evolved over the years.

NRM public policy initiatives were based on establishing NRM regions and developed an integrated regional NRM plan which recognized the interconnections of people to land, sea, coasts, fresh water systems and the biodiversity of the regions. Ecological Sustainable Development (ESD) formed the guiding philosophy underlying integrated natural resource management (INRM) plans. INRM advocated the adoption of the precautionary principle and followed adaptive management practices.

The INRM plan formed the basis for regional investments from both the National heritage trust as well as the National Action Plan for Salinity and Water Quality. These investments have facilitated a transition from the INRM towards

Table 1: Indigenous Ranger Groups in the Northern Territory (Source: NLC [12]).

Land Management or Ranger Group	Area
Acacia Larrakeyah	Adelaide River
Adjumarllarl Community	Gunbalanya
Asrikarrak Kirim	Peppimenarti
Bek & Mandiddi	W. Arnhem Land
Belyuen	Cox Peninsula
Demed	Gunbalanya
Dhimurru	Nhulunbuy
Djelk	Maningrida
Gamarrawa Nuwul Landcare	Nhulunbuy
Garawa	Borroloola
Garngi	Croker Island
Gumurr Marthakal Sea	Elcho Island
Gurrwurrling	Ramingining
Larrakia	Darwin
Lianthawirriyarra Sea	Gulf Carpentaria
Mardbalk Marine	Goulburn Island
Malak Malak	Daly River
Manwurrk	West Arnhem Land
Mimal	Bulman
Minyerri	Roper River
Mulayee Women	VRD
Ngaliwurru Wuli	Timber Creek
Ngatpuk	Finnis River
Numbulwar Numburindi Amalahgayag Inyung	Blue Mud Bay
SE Arafura Catchment	Central Arnhem Land
Thamarrurr Sea and Land	Wadeye
Waanyi Garawa	Boroloola area
Wagiman Guwardagun	Upper Daly River
Wanga Djakamirr	Ramingining
White Eagle	Finniss River
Yantjarrwu	Daly/ Pt Keats
Yugul Mangi	Ngukurr

integrated natural **cultural** resource management (INCRM). The conceptual template for NRM activities incorporated integration of the cultural elements that are deeply rooted in the psyche of the participants through the very process of acts performed by both the scientific community and the Aboriginal populations. The result of such actions can be seen when one recognizes that management of Aboriginal lands in the NT is currently through the Indigenous Ranger Groups.

Since Traditional Owners set up the Caring for Country program within the Northern Land Council in 1996, 35 Rangers Groups have been established which has employed over 400 Aboriginal people (Northern Land Council [12]). In many instances the Ranger Groups are given direction through an Aboriginal

Resource Centre which has a management committee made up of Traditional Owners from each clan in the management area. Since their initiation, Indigenous Ranger Programs have been considered as part of the Community Development Employment Program (CDEP) which is a 'work for the dole' type welfare program which has been in existence for 30 years in the NT and has both advocates (Altman [13, 14]); and critics (Pearson [15]). As such the Ranger Programs have been regarded as a CDEP activity which has been managed through a variety of different agencies. Funding for these Ranger Groups have come from variety of different funding bodies (National Heritage Trust, National Landcare Program, Indigenous Landcare Corporation etc.) and have often been dependent on the capacity of the Rangers or their coordinators to successfully bid for and win funding. Some groups have thrived under these arrangements and now have dozens of Rangers and projects while others have not had as much success and consist of only a handful of Rangers which provide few INCRM activities. Table 1 provides a list of INCRM activities in various areas of the NT.

4 Current status

In recent years PES has been a progressively larger source of funding for ranger activities as the main wages component has shifted from CDEP money. A major reason for such a significant adaptation of PES in NT is the attempt by both Commonwealth and the NT governments to increase market based activities.

Following Government Intervention in Aboriginal communities announced in July 2007, CDEP was planned to be abolished and changed over to a 'work for the dole' program. Although this full time employment in this program offered slightly more money ($20.80 per fortnight) there would be no opportunity for 'top-up', which is extra income for more than the 4 hours CDEP work per day (Australian Government [16]). With a change of Commonwealth Government in 2008 the plan to abolish CDEP was reviewed and it was recommended that CDEP be restructured. As of 1 July 2009 the 'continuing participants', or those already involved in a CDEP program, would keep receiving CDEP wages as long as they remained eligible. Those new participants who start in a CDEP program would receive income support from Centrelink which would be income managed. Centrelink is the Australian government authority responsible for providing access to government services, including social security allowances and employment plans. Furthermore, between 1 July 2011 and 30 September 2011 all of the 'continuing participants' would be moved off CDEP and onto income support (Australian Commonwealth Government [17]).

While these changes to CDEP were happening the Commonwealth Government put forward a *Working on Country* initiative was to fund the Indigenous Ranger Program. This was initially $20 million dollars over 5 years but was then expanded to $90 million over 5 years from 1 July 2008. This program now provides a wage to rangers which had superannuation, leave and sickness benefits that CDEP did not. It appears that formal INCRM is now following a much more reasonable employment path and these important activities area being recognized as a real job. There are however many Aboriginal people that are not receiving any benefit for the management services they are providing and it is this group who will struggle to remain on country with changes to CDEP.

In May 2009 the Northern Territory Government announced a new strategy and vision for Indigenous communities titled a *Working Future*. This strategy involved six elements which include: establishment of 20 towns across the Territory, a clear path for Outstations and Homelands, delivery of remote services in a targeted and coordinated way with the Australian Government, employment and economic development, a remote transport strategy and closing the Gap targets and evaluation (Northern Territory Government [18]).

The consequences of the *Working Future* strategy coupled with changes to CDEP are likely to contribute to urban drift of Aboriginal people from remote outstations and townships. A further depopulation of these remote areas will have long term biophysical and cultural impacts because it will disconnect people from country leaving less people involved in natural and cultural resource management. Indigenous Ranger Groups and the Indigenous Resource Centres that support Traditional Owners are looking for ways to generate an income from their country whilst also maintaining traditional knowledge.

5 Enhancing the future of NRM through PES

Attempts to enhance NRM by integrating the scientific and Indigenous knowledge systems have resulted in explicit recognition of the significant role played by the INCRM practices.

INCRM can be considered a major economic development opportunity for Aboriginal people seeing as they have knowledge, skills and capacity in this area. Given this opportunity, searches are underway to expand the INCRM practices to enable greater participation amongst Aboriginal people. Attempts are being made in earnest to drive the market based agenda based on INCRM. There are a number of PES activities that are already starting to provide an alternative avenue of funding which can support a presence on country. It is also necessary to recognize that some form of payment is rendered for the services provided by the Indigenous rangers. Table 2 summaries the characteristics of these PES activities in a form that is comparable with similar PES programs in other countries (adapted from Wunder [19]).

Existing PES schemes in the Northern Territory are still relatively new, but recent government initiatives have acknowledged the role of Indigenous Ranger Groups for services provided in relation to biodiversity protection, border protection, ongoing cultural maintenance, employment initiatives in remote communities and biosecurity (Muller [20]; Luckert et al. [21]). Many conservation and development organisations look at PES as a just way of rewarding poor rural people for looking after their environment which continuously provides environmental service (Shilling and Osha [22]; Rosa et al. [23]; van Noordwijk et al. [24]). However, others consider it from an efficiency point of view only as those that constitute a credible threat to ecosystem service

Table 2: Summary characteristics of PES programs carried out by Indigenous rangers groups.

Environmental service	Who buys	Who sells	Area
Disease monitoring -ants, mosquito, pigs, buffalo monitoring	Australian quarantine Inspection Service, Department of Agriculture, Fisheries and forestry	Adjumarllarl Community Rangers, Djelk Rangers, Larrakia Rangers, Mardbalk Rangers, Malak Malak people, Manwurrk Rangers, Ngaliwurru Wuli Land management, Yirralka Rangers, Yugal Mangi	Gunbalanya, Maningrida, Darwin, Goulbourn Island, Woolianna, Kabu warnamyo, Timber Creek, N.E. Arnhem Land, Ngukurr
Patrols for illegal, unreported and unregulated fishing	NT Fisheries	Mardbalk Rangers, Gumurr Marthakal Sea Rangers, Djelk Rangers, Lianthawirryarra Rangers, Thamarrurr Rangers, Tiwi Rangers, Anindilyakwa rangers, Yugal Mangi Landcare	Goulbourn Island, Elcho Island, Maningrida, Borroloola, Wadeye, Melville and Bathurst, Groote Island, Ngukurr
Illegal Fishing vessels	Australian Customs and immigration	Djelk Rangers	Arnhem Land Coastline
Ghost net surveillance management	Carpentaria Ghost Nets Programme	Dhimirru, Gumurr Marthakal Sea Rangers, Numbulwar Rangers, Yirralka, Yugal Mangi	N.E. Arnhem Land, Elcho Island, Roper River, Blue Mud Bay, Ngukurr
Monitoring and eradication of yellow crazy ant infestations	CSIRO	Dhimurru Rangers	East Arnhem Land
Landscaping and revegetation	Alcan Refinery, Nhulunbuy	Gamarrawa Nuwul Landcare	Nhulunbuy
Fire abatement	ConocoPhillips and NT Government	Jawoyn Rangers, Djelk Rangers, Demed Rangers, Manwurrk Rangers, Mimal Rangers	Maningrida - Katherine
Fire management	Neighbouring pastoral properties	Minyerri Rangers	Roper River

(ES) provision should be paid (Wunder [19]). This being the case, should only areas under immediate threat be open to PES activities? In practice there are no easy answers and PES schemes need to strike a balance between short-run efficiencies and long term viability (Wunder [19]). In the Northern Territory most Aboriginal land is under communal title and as such has restricted commercial development opportunity. Private investment is limited and land is generally agriculturally marginal but relatively intact providing functional ecosystem services. Government initiatives to boost the economy on Aboriginal land through encouraging private investment (by providing greater long term security through allowing for 99 year leases and abolishment of the permit system) along with greater activity from the mining industry from a recent resources boom are contributing to an increase in threats to ecosystem services on this country.

Indigenous people are well equipped to provide valuable input into the management of resources and associated ES under a PES type agreement due to their considerable existing knowledge and skills base, demonstrated commitment and location. However, most of the activities that Indigenous rangers provide cannot be strictly classified as PES activities as defined by Wunder [19] as although their ICNRM activities may maintain a variety of ES their payment may not necessarily be based on performance. There are other funding models such as the now defunct Top End Aboriginal Land Management and Employment Strategy which was funded largely to control the spread of an exotic plant *Mimosa pigra* (Ashley et al. [25]) which could be considered similar to a PES activity as performance was an important criteria for continuation of funding. There are however a number of activities that are currently operating through Indigenous Ranger Groups in the Northern Territory that can be classified as PES activities (Table 3). The majority of these schemes are

Table 3: Summary of the Indigenous land and Sea Rangers manager activities that can be classed as PES activities.

Ranger Activity	Fire Management	Weed Control	Feral animal control	AQIS - mosquito/blood	Coastal Management	Coastal surveillance	Illegal foreign fishing vessel surveillance	Ghost net surveillance	Feral ant management	Land Rehab	Cultural site management	Fauna and flora surveys	Junior ranger program
PES Activity	√	√	√	√	√	√	√	√	√	√	√	√	√

government financed programs but there are also a few user-financed programs involved. These activities are voluntary with the ES generally being well defined but, as found by Wunder [26] it appears that in most cases payments are nominally conditional within the government-financed programs than in user-financed programs (such as WALFA, exotic ant control).

Some have suggested that the Australian Government could be considered to be 'free-riding' on the provision of these INCRM services by Aboriginal people (Luckert and Whitehead [27]). These management practices deliver outputs which are of public good/benefit (reduced carbon, disease control, coastal surveillance, feral animal and weed control etc) at several levels (local, national and international) (Muller [20]; Luckert et al. [21]). As such, all of the people that provide their services should be recompensed for their activity. Notwithstanding this, the basic premise has been to provide a meaningful way for Aboriginal people to be gainfully engaged in promoting management of the five capitals – economic, environment, social, cultural and human, in a sustainable manner.

The future of PES in enhancing NRM will largely depend upon the extent of involvement of Indigenous people to the market based agenda. Records based on past experiences with employment, education and enterprise development initiatives that are predominately market oriented are not very encouraging. A major reason for such a poor track record of market engagement by Aboriginal populations is the adoption of a 'deficit' approach to private sector involvement.

PES has all the essential in-built features to become an imaginative tool for enhancing NRM practices. To achieve its full potential there is a need to take heed of lessons from current and past experiences.

There are a number of PES activities happening in the Northern Territory that are providing employment and income generating opportunities for Indigenous Ranger Groups that have the capacity and skills for these activities. Many of these activities operate through Ranger Groups because transaction costs are lower and these groups have a certain amount of operational capacity. However, there are many Indigenous people living on Aboriginal land in remote parts of the Northern Territory who are not affiliated with a Ranger Group, who also have NCRM skills and obligations, who are reliant on CDEP and would welcome the opportunity to participate in PES activities. It is important to develop a framework which allows participation from a larger pool of people across the NT than just those affiliated with Ranger Groups.

Experiences from other similar environments such as Canada suggest the significance of broader involvement.

> "Broader involvement at the community level was restricted because specific local actors coalesced to skew participation toward a narrow set of economic values. Selected participants in the co-management exercise maintained close control over participation within the community and effectively became a conduit to senior government for vocal sectors at the local level. The locally

> dependent status of committee members resulted in proposals for modest change, without an overall vision or commitment to implementation."

A starting point in embracing such a lateral approach is to recognise that provision of environmental services is neither about science nor service provision alone but about context, culture (Muller [20]) connection, continuity, change, competition, and cooperation. Thus, there is potential to make the process that of 'a-colonisation' where Indigenous and non-Indigenous governance systems connect and work together rather than a top down or prescribed 'colonised' approach. In addition to creating a more correct process of engagement, recognition and remuneration for environmental services, it is essential to conceptualise how INCRM services can be assigned appropriate values, promote livelihood benefits, and that can be distributed to all the people involved in providing these services. In short, it is not simply generating processes of production by Aboriginals themselves for themselves and the rest of the Australian society but it involves adoption and promotion of a 'holistic' perspective that includes production, distribution and consumption values of environmental services.

PES should therefore not be restricted to becoming simply a tool for calculating monetary values of payment for services rendered (resulting from market failure) but it should be able to incorporate how services that are produced, consumed and distributed in a community setting can be valued. As a result PES becomes an organizational framework that can be used effectively to manage the environment by espousing sustainable management principles. There are several initiatives that are built on market based instrumentations related to price; quantity and market features should be incorporated when discussing PES. The PES framework then can generate initiatives that are based on a premise to find alternatives to publicly funded operations. However this has not happened to a large extent in the case of Northern Australia.

References

[1] Hicks, J. *A theory of economic history,* Claredon Press, p.3, 1969.

[2] Hodgson, G.M., *The economics of Institutions,* International library of critical writings in economics, **33**, Elgar Reference Collection, 1993.

[3] de Groot, R. *Functions of Nature: Evaluation of Nature in Environmental Planning, Management and Decision Making*, Wolters-Noordhoff, Groningen, the Netherlands, 1993.

[4] Rose, D.B., *Nourishing terrains: Australian Aboriginal views of landscape and wilderness*, Australian Heritage Commission, Canberra, 1996.

[5] Altman, J.C., Kerins, S., Fogarty, B. & Webb, K., *Why the Northern Territory government needs to support outstations/homelands in the Aboriginal Northern territory and national interests* , Centre Aboriginal Economic Policy Research Tropical Issue No. 17/2008, Centre for

Aboriginal Economic Policy Research, Australian National University, Canberra, 2008.

[6] Department of Natural Resources the Arts and Sport, *Northern Territory Parks and Conservation Masterplan*, Northern Territory Government, Darwin, 2005.

[7] Garnett, S.T. & Woinarski, J.C.Z., A case for Indigenous threatened species management (Chapter 5). *Investing in Indigenous Natural Resource Management, ed.* M.K. Luckert, B.M. Campbell, J.T. Gorman and S.T. Garnett, pp. 38-44, Charles Darwin University Press, Darwin, 2007.

[8] SRRATRC. *Commercial utilisation of Australian native wildlife,* Report of the Senate Rural and Regional Affairs and Transport References Committee, Parliament of the Commonwealth of Australia, Canberra, 1998.

[9] Handy, C., *The Elephant and the Flea*, Random House, Arrow, NSW, 2002.

[10] Altman, J.C., *Sustainable development options on Aboriginal land: The hybrid economy in the twenty-first century*, Centre Aboriginal Economic Policy Research Discussion Paper 226, Australian National University, Canberra, 2001.

[11] Pearson, D.M. & Gorman, J.T. Managing landscapes of the Australian Northern Territory for sustainability: Visions, issues and strategies for successful planning, *Futures*, 2009 (in press).

[12] Northern Land Council, *Celebrating Ten Years of Caring for Country, a Northern Land Council Initiative*, Northern Land Council, Darwin, 2006.

[13] Altman, J.C. *In Search of an Outstations Policy for Indigenous Australians*, Centre Aboriginal Economic Policy Research, Australian National University, Canberra, 2006.

[14] Altman, J.C. Scrapping CDEP is just plain dumb. *Crickey Magazine*, 2007.

[15] Pearson, N. Stuck on the Welfare Pedestal. *The Weekend Australian: Inquirer*, p. 28, Perth, 2007.

[16] Australian Government, *Community Development Employment Projects (CDEP) Program. Program Guidelines 2009-12*, Department of Families, Housing, Community Services and Indigenous Affairs, Canberra, 2007.

[17] Commonwealth Government of Australia, *Increasing Indigenous economic opportunity*. A discussion paper on the future of the CDEP and indigenous employment programs Department of Education, Employment and Workplace Relations, Canberra, 2009.

[18] Northern Territory Government. *Working Future*. http://www.working future.nt.gov.au/. Canberra, 2009.

[19] Wunder, S., *Payments for Environmental Services: some nuts and bolts*, Occasional Paper No. 42, Centre for International Forestry Research, Bogor, 2005.

[20] Muller, S. Indigenous Payment for Environmental Service (PES) Opportunities in the Northern Territory: negotiating with customs, *Australian Geographer* **39**, pp. 149-170, 2008.

[21] Luckert, M.K., Campbell, B.M., Gorman, J.T. & Garnett, S.T. (eds.). *Investing in Indigenous Natural Resource Management*, Charles Darwin University Press, Darwin, 2007.
[22] Shilling, J. & Osha, J., Paying for environmental stewardship: using markets and common-pool property to reduce rural poverty while enhancing conservation, World Wildlife Fund, Washington D.C., 2003.
[23] Rosa, H., Kandel, S. & Dimas, L., Compensation for environmental services and rural communities, Programa Salvadoreño de Investigación sobre Desarrollo y Medio Ambiente (PRISMA), San Salvador, El Salvador, 2003.
[24] van Noordwijk, M., Chandler, F. & Tomich, T.P., An introduction to the conceptual basis of RUPES: rewarding upland poor for environmental services, World Agroforestry Center, Bogor, 2004.
[25] Ashley, M., Storrs, M.J. & Brown, M., Caring for Country: Community based mimosa management on Aboriginal lands of the Top End of the Northern Territory. *Proceedings of the 3rd International Mimosa Symposium*, pp. 106-109, CSIRO, Darwin. (2002)
[26] Wunder, S. Necessary conditions for ecosystem service payments *Proceeding from Economics and Conservation in the Tropics: A Strategic Dialogue*, San Francisco, 2008.
[27] Luckert, M.K. & Whitehead, P.J. A general case for natural resources management: market failures and government policy (Chapter 2). *Investing in Indigenous natural resource management*, ed. M.K. Luckert, B.M. Campbell, J.T. Gorman and S.T. Garnett, Eds.), pp. 11-18, Charles Darwin University Press, Darwin. 2007.

Investment in sustainable buildings: the role of green building assessment systems in real estate valuation

S. Geissler & M. Groß
Department of Buildings & Heating, Austrian Energy Agency, Austria

Abstract

In Austria, the building sector accounts for more than one-third of energy consumption and material flows. There are several instruments to improve building quality in terms of environmental performance, such as building codes, subsidies, and voluntary building assessment schemes. In this paper, the focus is on mandatory and voluntary building assessment schemes and their role in real estate valuation. Real estate valuation is based on calculation methods and on the assessment of market demand. Therefore, if there is a demand for energy efficient and sustainable buildings on the market, high performance buildings will be better valued than average buildings. Building assessment schemes serve to create awareness for the benefits of high performance buildings, thus increasing the market demand. However, work is also done on further developing real estate valuation calculation methods, in order to identify risks for future valuations that might be connected with, for example, the energy consumption and material properties of the respective building. This paper presents the findings of a project on further developing real estate valuation methods by taking environmental performance into account. It describes the voluntary building assessment schemes and the mandatory energy certification scheme applied in Austria, and then puts focus on the guideline for appraisers developed in the project. The guideline provides support on how to consider energy aspects and other sustainability aspects in building valuation.

Keywords: Austrian green building assessment scheme, TQB assessment, real estate valuation.

WIT Transactions on Ecology and the Environment, Vol 131, © 2010 WIT Press
www.witpress.com, ISSN 1743-3541 (on-line)
doi:10.2495/EEIA100161

1 Introduction

In Austria, the building sector accounts for more than one-third of energy consumption and material flows. Building assessment schemes are important drivers for increasing building quality and reducing impacts on the environment at the same time. This paper tackles their role in real estate valuation and answers the question how real estate valuation can be used as an instrument for increasing building quality in terms of environmental performance.

Real estate valuation has a long tradition, and comes in as soon as a transaction is planned. It assesses a site including the building in monetary terms, and the assessment result is the value in a defined currency. The location of the site (e.g. connection with public transport, status) and the building use (residential, non-residential, industry, tourism) have both been dominating assessment results so far, because risk concerning the income generated by renting or selling the building was seen as mainly connected with these two criteria.

Real estate valuation is based on experience and knowledge about market demand and willingness to pay for a certain type of building located on a specific type of site. It is also based on rules of thumb concerning the estimation of cost parameters and risks. The procedure has to be short due to cost constraints for valuation certificates, especially in the residential sector. There is hardly time for detailed building specific data surveys, and appraisers often do not have the technical expertise for assessing the technical quality of a building. Usually, the educational background of appraisers is law or economics, and if a specific technical expertise is needed, an expert's opinion is commissioned.

By definition, the major influence on the real estate valuation result is market demand and willingness to pay due to the methodology of valuation. Therefore, the appraiser will not account for investments in energy efficiency measures in the valuation report if the market does not ask for energy efficient buildings, and does not reward energy efficiency in terms of willingness to accept higher prices for renting or buying. On the other hand, if investment in energy efficiency and other quality improvements is not rewarded on the market, the majority of developers and building owners will not go for sustainable buildings. This traps us in the "the chicken or the egg" causality dilemma. The Energy Performance Buildings Directive (EPBD) 2002/91/EC addresses this situation, aiming towards transforming the real estate sector towards more energy efficiency by regulating that an energy certificate must be issued for each building or unit of use [1]. It is the objective to change consumers' demands by making the energy efficiency of a building transparent. However, due to the way of national implementation in Austria, the EPBD is less effective than it could have been. In fact, many buildings do not have an energy certificate because regarding the building stock the national law allows for exemptions for all building types. Nevertheless, there is a certain dynamism, also due to discussions on climate change and security of energy supply, and especially in connection with the spreading of voluntary green building assessment schemes covering aspects above and beyond energy.

Green building assessment schemes have been developed to display the environmental quality of a building and to raise demand for so-called green buildings. Assessment results of green building schemes are expressed in the form of a label. The objective is to make the benefits of the building transparent, such as reduction in energy cost during building operation, but also other non-monetary benefits such as the increase in comfort or positive effects of good indoor air quality on the health of occupants. Of course, these effects should also result in the willingness to pay more for a building, increasing the building's value.

Developers as well as appraisers and other real estate experts have realised that energy but also other sustainability aspects are gaining in importance. This has raised interest from many important stakeholders in tackling the question of how to deal with energy efficiency and other sustainability issues in real estate valuation.

2 Voluntary green building assessment schemes

The first green building assessment scheme BREEAM (BRE Environmental Assessment Method) was developed in UK and has been applied since the 1990s. The objective was to demonstrate that green buildings have an edge over conventional ones and that "green" pays off due to better comfort, reduced energy costs and less harmful impact on the environment. In the 1990s the development of national green building assessment schemes took place in many countries, also supported by the international initiative "Green Building Challenge" led by Canada [2]. The British BREEAM and the LEED scheme developed in the USA are also offered at the international level.

According to RICS, a green building is a property that uses resources efficiently, reduces waste and provides superior indoor air and other qualities [3]. Recent market analysis shows that green buildings (office buildings as well as residential buildings) increase in value compared with conventional ones [3–5].

All building assessment schemes consist of the following elements:

(1) Assessment system

- Criteria: which kind of information is needed for assessment (e.g. heating energy consumption)
- Indicators: how to describe the performance of the defined criteria (e.g. kilowatt hours per square meter heated gross area and year)
- Assessment scale: defines which performance is good and which one is bad by allocation of scores (e.g. heating energy consumption less than 15 kilowatt hours per square meter heated gross area and year receives the highest score)
- Weighting: which criteria are more important than others, and by how much (e.g. more points are allocated to energy related criteria than to material related criteria)

(2) Assessment procedure

- Data collection
- Check of data

- Awarding of points based on the data provided
- Awarding the certificate based on the assessment result

The assessment result contains two parts:

- Compilation of quantitative data and qualitative information about the building.
- Assessment result for market communication, in order to tell consumers how good the performance of the building is.

Data collection and data assessment should be well separated, to make use of the data apart from the green assessment scheme. The data collection contains objective information about the building, which is very useful for other purposes, among others for the operation of the building. While the data compilation as such remains the same, the interpretation may change, depending on the assessment scheme applied.

Looking at the elements of a building assessment scheme, it becomes evident that voluntary schemes can be used for several purposes.

(1) On the company level:

- To use the assessment criteria for revising the design targets in order to go for a high quality building
- To use the assessment scheme for quality control
- To assess the quality of a building and to use the assessment result in market communication

(2) On the political level:

- To contribute to the transition of the construction sector towards sustainability by
    - creating awareness on building quality in order to raise demand for high quality buildings in terms of reduced resource consumption;
    - using voluntary schemes in order to prepare the construction sector for mandatory regulations to come.

3 Building assessment schemes in Austria

In cooperation with the Green Building Challenge the Total Quality assessment scheme was developed and after a testing phase went into operation in 2003 [6]. It was decided to elaborate a national assessment scheme, based on international experience, but adjusted to the Austrian planning and construction practice, in order to avoid high transaction costs. In Austria, small and medium-sized companies are in the majority, and it was the objective to provide a cost-efficient tool for widespread application. In 2009 the system was revised based on lessons learnt and renamed as Total Quality Building (TQB) assessment scheme.

Although the assessment scheme was developed according to developers' needs, market up-take was quite slow, and especially the use of the assessment results in marketing communication was lacking. It turned out that vendors did not know how to use the assessment result in the selling or renting process, largely because they were not trained in aspects such as indoor air quality, primary energy consumption of materials or CO_2-emissions. It seemed as if the barrier for the construction sector was too high, with exception of a few innovators. Therefore, green building assessment was integrated in the climate protection programme “klima:aktiv”, launched by the Austrian Federal Ministry of Agriculture, Forestry, Environment and Water Management in 2004. The programme targets a substantial reduction of CO_2-emissions and a substantial increase in energy efficiency in the building and transport sector, in industry, and households, by funding comprehensive activities such as the production of information material, consulting services, network development, and the elaboration of quality control procedures [7].

The sub-programme “construction and refurbishment” offers the klima:aktiv building standard for residential and non-residential buildings, and for new build constructions as well as for refurbishments. This standard consists of selected criteria of the TQB standard with focus on outdoor and indoor environment, to communicate individual benefits along with the reduction of CO_2-emissions.

The criteria system comprises the following categories:

- A Design and Construction
- B Energy and Supply
- C Materials and Structure
- D Comfort and Indoor Air Quality

To lower the barrier for potential applicants of the building assessment scheme, and to make use of the dynamism stemming from the EPBD, a step-wise system has been introduced since the full implementation of this EU-Directive in 2009: it consists of the mandatory energy certificate according to EPBD representing the first step and the core at the same time. According to EPBD, buildings have to meet minimum requirements in terms of energy consumption, which fully correspond with the voluntary klima:aktiv assessment category “B energy and supply”. Documenting the fulfilment of additional criteria concerning “A Design and Construction”, “C Materials and Structure”, and “D Comfort and Indoor Air Quality” results in the klima:aktiv building award. Meeting additional criteria concerning noise protection, flexibility, and others, and having all data checked by authorised staff results in the TQB certificate. In this way, companies have the chance to become familiar with environmentally conscious design and construction, and concerning practical work the procedure is effective and time saving.

klima:aktiv was developed based on a study commissioned by the Austrian Federal Ministry of Agriculture, Forestry, Environment and Water Management, and TQB was developed based on studies commissioned by the same ministry, the Austrian Federal Ministry for Transport, Innovation and Technology, and the Austrian Federal Ministry of Economy, Family and Youth. Since 2009, the ÖGNB - Austrian Council for Sustainable Construction - has been running both building assessment schemes [8].

4 Sustainable building quality in real estate valuation

In 2008, an applied research project started, aimed at the development of a guideline on how to deal with energy efficiency and other sustainability aspects in real estate valuation. The project was carried out at the national level in Austria, it was initiated by the Austrian Energy Agency, and was funded by the Austrian Climate and Energy Fund. A team consisting of real estate experts and experts in the field of sustainable construction joined forces in this project. In addition to the core project team, a group of experts representing all stakeholder groups from the real estate valuation sector participated in workshops to discuss crucial issues and ensure that project results will be useful for practicing appraisers.

In the starting phase of the project, a workshop was organised with the participation of the core project team and the valuation experts. The group discussed (1) energy efficiency and sustainability in the building sector in general, (2) the potential role of energy certificates issued according to EPBD and the role of voluntary green building certificates such as the Austrian TQB system, and (3) the criteria building assessment schemes have to fulfil to produce certificates with high acceptance on the market.

The group agreed to focus on the role of energy in real estate valuation, because there is an existing legal basis for this aspect exclusively, represented by the EPBD. Other sustainability aspects covered by voluntary assessment schemes were seen as less important, because the number of buildings endowed with a certificate will be limited as long as the scheme is not mandatory. Furthermore, energy certificates provide the information which is necessary to calculate energy cost savings or additional expenses for energy compared with a reference building, and this approach was used to incorporate energy efficiency in monetary terms.

4.1 Methodological approach

Real estate valuation is based almost without exception on experts' know-how about market demand and, according to the Austrian standard ÖNORM B 1802, has to be based on methods which meet the up-to-date scientific standards. The project team focused on the income approach to valuation, which is referred to in the Austrian standard mentioned above as one of the applicable methods, and which is widely used in Austria. The method seemed to be appropriate for investigating the role of energy efficiency and other sustainability issues in real estate valuation. To explain it in a simplified way, this method sums up all income potentially generated by rent for a defined period and subtracts investor's operating expenses. This determines the net operating income over the useful life of the property, and then a discount rate is used to calculate the present value, which is used to evaluate the potential for investment.

5 Project results

5.1 Three options to consider energy efficiency, and one to be recommended

During the course of the project, several methods to consider energy efficiency in the valuation results were tested by calculating 20 multi-unit residential buildings and 15 office buildings. Finally, three options were determined. Using the income approach to valuation, energy efficiency can be considered when calculating the annual net operating income, or when choosing the discount rate or when calculating the estimated building value:

- Option 1: Calculation of annual net operating income: energy efficient buildings reduce energy costs during operation. According to the total cost of ownership approach, customers will be ready to pay higher rent if they save on operating costs. Therefore, the calculation considers a higher potential gross income based on rent which increases for the amount corresponding with the amount of saved energy cost, compared to a reference building.
- Option 2: Discount rate: Energy efficiency reduces future exploitation risks. Buildings with a poor energy performance are exposed to the risk of potentially increasing energy prices. Demand on the market will drop for buildings with high energy consumption. Therefore, it is possible to consider this future risk connected with the energy efficiency of a building with a surcharge on or a deduction from the discount rate.
- Option 3: Calculation of the estimated building value: Possible extra expenses for energy compared with a reference building and as a consequence of bad energetic building quality pose a risk for future lettability or saleability of the property, or the contrary in case of energy cost savings. Therefore, possible extra expenses for energy or savings compared with a reference building are considered when the estimated building value is calculated. The heating energy consumption data of the specific object and the reference building are used to determine the possible extra expenses or savings. A specific line is added to the calculation procedure to make sure that energy efficiency is taken into account.

The project team found option 3 to be recommendable for practicing appraisers, for the following reason: The introduction of a defined step in the calculation procedure for the consideration of energy efficiency ensures that that there is no double counting on the one hand, and no omitting on the other hand. For instance, energy efficiency and the associated risk concerning lettability can be also taken into account when determining the loss of rental income risk, which is a step in calculating the estimated building value. However, in the valuation report it is not transparent whether the appraiser has considered energy efficiency or not. The same applies for the determination of the discount rate, and the calculation of the annual net operating income. Therefore, highlighting a specific step in the calculation procedure of the income approach to valuation ensures that energy efficiency is either taken into account, or is consciously not taken into account, because it is not relevant for the type of real estate property under assessment. The project team and the experts' group assessed this solution also to be especially useful concerning awareness creation about energy efficiency, which is still lacking in the real estate sector.

Test calculations showed that this option generates real estate values, which differ up to 10% from the value calculated without taking energy efficiency into account, depending on the energetic performance of the building and depending on the reference building chosen.

5.2 The reference building

The calculation of extra expenses or savings on energy is carried out in comparison with a reference building with a specific amount of energy consumption for space heating. Domestic hot water consumption is not considered because it depends less on the quality of the building than on the user behaviour. In Austria, there are two options to determine the reference for heating energy consumption.

- Building code: in the course of implementing the EPBD in Austria, building codes had to be revised to show those heating energy consumption values which represent the minimum performance to be achieved at the least [9].
- Agreement according to Article 15a Constitutional Law between the federal state and the provinces: this Austrian policy instrument addresses the social housing subsidy schemes and public non-residential buildings by giving a limit of heating energy consumption to be achieved at the least [10].

Heating energy consumption numbers in the Article 15a Agreement are more ambitious than those in the building code. Usually, the building codes follow the specifications of the Article 15a Agreement after a certain period of time. Choosing the more ambitious reference building (according to Article 15a Agreement) leads to a stronger decrease in value of less energy efficient buildings than if this was the case with the reference building according to building code. The discussion about the choice of reference building can be summarised as the question of whether there should be an increase in value of energy efficient buildings or a decrease in value for buildings lacking energy efficiency. While the non-residential sector prefers to reward energy efficiency, the residential sector is concerned about increasing costs for occupants and the consequential social problems, and prefers a decrease in value for buildings with a poor energy performance.

5.3 Extended valuation report

In contrast to energy efficiency, it is much more difficult to describe other sustainability aspects in monetary terms, such as use of daylight and healthy indoor air quality which result in increased wellbeing and greater productivity when dealing with office buildings. Apart from the assessment result, green building assessment schemes generate a lot of objective information about a building (see chapter 4), for instance on daylight, materials, indoor air pollution, flexibility, noise, safety and security. Part of this information can be used to assess possible risks if appropriate for the property under assessment. Among others, the project resulted in the strong recommendation to extend the valuation report and include a risk assessment based on information from green building assessment certificates.

5.4 The guideline for appraisers

The project resulted in a guideline which contains a description of the relevance of energy efficiency and other sustainability aspects. The focus is on options of how to consider energy efficiency in the income approach to valuation. Option 3 Calculation of the estimated building value (see description above) is explained in detail, and all parameters for calculation are described. The guideline leaves the decision to the appraiser regarding whether to take into account energy efficiency or not, because there will be real estate properties where energy efficiency is not relevant at all due to the special location of the building or due to a special type of use. However, it contains useful information on energy consumption of typical buildings by construction period with and without energetic refurbishment measures, and explains the energy certificate according to EPBD as data source for calculating extra expenses or savings on energy. It also describes green building assessment schemes, and the information they contain which might be useful for appraisers.

6 Conclusions

The income approach to valuation demands that valuation results refer to a defined key date. Nevertheless there is also the rule that already known aspects, which might influence the value of the property in the future, have to be taken into account. This is the justification to deal with the role of energy efficiency in real estate valuation on the one hand, but also to deal with other sustainability issues on the other hand, looking at the increasing importance of voluntary green building assessment schemes, e.g. in connection with real estate investment funds. It is evident that the real estate sector is already in the process of changing: Surveys conducted among other countries in USA show that market demand for energy efficient buildings exists and that transactions are based on higher prices compared with conventional buildings [11].

In Austria, there are clients who explicitly ask for the consideration of energy efficiency when they commission a valuation report. Therefore, the professional profile of appraisers is also changing and is becoming more complex due to the ongoing technical progress in the building sector and specifications on energy efficiency and efficient resource utilisation in general given by the respective EU regulations. This has to be considered in the further development of education and training offers.

The energy certificate according to EPBD provides data for the valuation procedure and thus plays an important role. Unfortunately, in Austria acceptance in the real estate sector is quite low at present due to the missing quality of energy certificates. However, with the EPBD Recast there will be the obligation to establish a quality control system for energy certificates combined with the requirement that only qualified personnel will be allowed to issue them [12]. This will facilitate acceptance in the real estate sector and contribute to making use of the respective data in the valuation procedure.

Due to the quite ambitious building codes implemented as a consequence of the EPBD in 2008, energy efficiency will soon be standard in Austria. Therefore, existing buildings without energy efficiency refurbishment will decrease in value. Comprehensive green building assessment schemes, which are applied on a voluntary basis, also consider impacts of materials, indoor air quality and other aspects. Therefore, building quality is constantly improving due to awareness creation and market demand raised for sustainable buildings. However, this process is expected to move slowly compared with the dynamism in the field of energy. Here, it became evident that the market only solves the problem if the correct framework conditions are given. The revised version of the EPBD imposes stronger conditions on the building sector. It tackles energy efficiency but also associated aspects such as indoor climatic conditions. Therefore, another building quality aspect will soon become mandatory and which will contribute to further transforming the building sector.

References

[1] Directive 2002/91/EC of the European Parliament and of the Council of 16 December 2002 on the energy performance of buildings (EPBD). Official Journal of the European Communities, 4.1.2003

[2] Howard, N. (2006): Building Environmental Assessment Methods: in Practice. In: The 2005 World Sustainable Building Conference, Tokyo, 27.–29. September 2005, Conference Proceedings: 2008-2015

[3] Royal Institution of Chartered Surveyors (2005): Green Values – Green buildings, growing assets. London: RICS

[4] Eichholtz, P.; Kok N.; Quigley J.M. (2009): Doing Well by doing Good? Green Office Buildings. Working Paper No W08-001; Fisher Center for real Estate and Urban Economics, University of California, Berkeley, January 2009

[5] Salvi, M.; Horejájová A.; Müri R. (2008): Minergie macht sich bezahlt, Erika Meins (Hrsg.). Zürich: CCRS und Zürcher Kantonalbank

[6] Geissler, S.; Bruck, M.; Lechner, R. (2004): Total Quality (TQ) Planung und Bewertung von Gebäuden. Wien: Berichte aus Energie- & Umweltforschung 08/2004, Bundesministerium für Verkehr, Innovation und Technologie.

[7] Austrian climate protection programme, http://www.klimaaktiv.at/article/archive/11911/

[8] ÖGNB, TQ and TQB certified buildings, http://www.oegnb.net

[9] OIB Richtlinie 6 Energieeinsparung und Wärmeschutz, Ausgabe: April 2007. Wien: Österreichisches Institut für Bautechnik.

[10] Vereinbarung gemäß Art. 15a B-VG zwischen dem Bund und den Ländern über Maßnahmen im Gebäudesektor zum Zweck der Reduktion des Ausstoßes an Treibhausgasen. http://www.wien.gv.at/recht/landes recht-wien/landesgesetz blatt/jahrgang/2009/pdf/lg2009045.pdf

[11] Bienert, S.; Steixner, D.; Koch, D. (2009): Integration of energy efficiency and LCC into property valuation practise. Transforming green features into values. Paper for the Pacific RIM Real Estate Society 15th annual conference, 18-21 January 2009, Sidney
[12] Agreement on EPBD Recast 18. November 2009, http://www.eceee.org/buildings/EPBD_Recast/

Hydropower and sustainable development: a case study of Lao PDR

S. Jusi
Finland Futures Research Centre, Turku School of Economics, Finland

Abstract

Lao People's Democratic Republic (PDR) has large hydro-power potential with estimates varying from 18,000 MW to 30,000 MW, out of which only two percent (667,5 MW) has been developed so far. Despite the fact that the hydropower sector has played a pivotal role in the economic development of Lao PDR, planning, selection and implementation processes for hydropower projects have tended to be ad hoc in character, have experienced insufficient transparency and have not delivered the full potential benefits to the development of the country. Whereas substantial improvements in policies, legal requirements and assessment guidelines have occurred, hydropower planning and development still need a lot of improvement. Some major problems relate to capacity and institutional environment, such as insufficient quality of environmental and social assessments, ineffective regulatory framework, a lack of transparency, and the failure to conduct comprehensive consultations with all stakeholders.

Hydropower development in Lao PDR and in the whole Mekong Region needs to be sustainable, which will require minimal adverse social and environmental impacts, while remaining a viable, profitable and source of renewable energy supporting the country's economic development. Hydropower development should also better respond to the realities of regional energy market. Hydropower must be developed cautiously in the context of broader development goals, including responsible environmental management, institutional development, poverty alleviation and social development along with integrated water and energy management. Sustainable development of hydropower requires the integration of economic and social development and environmental protection and the need to take into account the values of efficiency, participatory decision-making, sustainability, accountability along with precautionary approach to environmental management and eco-efficiency. The purpose of the paper is to identify key challenges related to sustainable hydropower planning and development in Lao PDR. The paper aims to

WIT Transactions on Ecology and the Environment, Vol 131, © 2010 WIT Press
www.witpress.com, ISSN 1743-3541 (on-line)
doi:10.2495/EEIA100171

contribute to the broader discourse on sustainable hydropower planning and development in developing countries.

Keywords: hydropower, Lao PDR, sustainable, development, planning, risks.

1 Introduction and background

Management of water resources in an effective, efficient, equitable and sustainable manner will be essential for ongoing national level. The national objectives of the hydropower development are to expand an affordable, reliable and sustainable electricity supply, to promote economic and social development and to overcome Lao PDR's comparative disadvantages in attracting industry and investment [29, 43]. The Government of Lao PDR (GOL) has committed itself to increasing electrification to 90 percent by 2020 [29, 35]. Hydropower generation is also an opportunity for the country to contribute to international efforts to reduce greenhouse emissions.

There are high potentials of electricity that can be generated from about six to eight small dams, which are capable of generating 600–800 MW [36]. The construction of some other major power plants/projects as part of the five-year plan has commenced, including the Nam Theun 2 Hydropower Project, which will add another 1,070 MW capacity by the end of 2009. The GOL has to date signed MOUs or is undertaking research studies on a total of more than 70 hydropower projects [11]. On average 65 to 80 percent of the annually produced energy is exported [48].

The amount of power imported is expected to grow as a result of the rapidly increasing demand and the constraints on power generation [33]. Currently, domestic energy consumption is growing at eight to ten percent annually [48]. Until the year 2020, the electricity demand forecast is 7,770.7 GWh (average growth rate is 13% per year) and peak load is 1,486.8 MW (average growth rate is 11% per year) [12]. Due to its energy surplus and geographical location at the hub of the Greater Mekong Sub-region, the Lao PDR is strategically positioned to play a significant role in promoting regional power trade [29, 31, 44].

Currently the Lower Mekong Basin (LMB) countries are net importers of fossil fuels, therefore hydropower projects would bring significant macro-economic benefits by reducing dependency on energy imports from outside the region. Measures to combat climate change may also be expected to include a greater dependence on hydropower as a renewable source of electricity and raises a major sustainability issue of global importance [8]. The issue of climate change and hydropower is important one but not discussed here due to limitation of space (see more for example [1–3, 10, 21, 46]).

Hydropower investment decisions by the public and private sectors alike have been made on an individual project basis without reference to any plan to ensure that the priority and scope of projects are consistent with optimal development objectives. When developed appropriately, with attention to social and environmental risks, hydropower projects can provide multiple economic and environmental benefits to the poor people of Lao PDR. The literature on sustainable approach in hydropower development [26, 46, 47] emphasizes the importance of different aspects of sustainability in the successful implementation of the sustainable approach. These aspects are social, economic and environmental sustainability together with precautionary and eco-efficiency approaches. The key question to consider is how to provide energy in a way that is consistent with the goals and principles of sustainable development, ensuring a better quality of life for all people, including future generations [30].

This paper is not intended to be a comprehensive analysis, but rather a ‘discussion startup’ that will help steer both hydropower/energy and environment decision-makers as well as future work to strengthen the sustainable and sound environmental, social and economic development of Lao PDR.

2 Hydropower and sustainable development context

Hydropower has been perceived and promoted as a comparatively clean, low-cost, renewable source of energy that relies on proven technology [46]. Except for reservoir evaporation, which makes the hydropower to have a big water footprint [17] it is a non-consumptive use of water. At the same time, the hydropower can also be viewed as large ‘ecological footprint’ sectors – that is, it has substantial impacts on the natural environment because of its use and reliance upon natural resources [10]. Also, classified as a clean, renewable energy source, hydropower can reduce the net production of greenhouse gases by moving economies to a lower-carbon future by displacing other forms of power generation [20, 47].

There are risks inherent in the development and operation of hydropower. These risks cross the range of financial, engineering, geological, and market concerns, with particular attention to environmental protection, social inclusion, resettlement and sharing of the benefits of development across all stakeholders. Hydropower reservoirs have a large number of potential cross-sectoral impacts, including changes in downstream flows and water quality, dam safety, in-stream and reservoir fisheries, resettlement, ecological impacts, and flood control [10, 19, 44, 46]. As a consequence of a wide variety of impacts and risks, the definition of acceptable hydropower has shifted to one that recognizes the core principles of sustainable development, with attention to social, environmental and economic aspects [8, 27, 47, 50].

2.1 Sustainable development

The Brundtland Commission has defined sustainable development as being development that meets the needs of the present without compromising the ability of future generations to meet their own needs [40]. Sustainable development requires the integration of three components – economic development, social development and environmental protection – as interdependent, mutually reinforcing pillars. Eradicating poverty, changing unsustainable patterns of production and consumption, and protecting and managing the natural resource base underpinning economic and social development are overarching objectives of, and essential requirements for sustainable development [50].

International organizations like the International Hydropower Association [27] and World Commission on Dams [46] among the international financiers [47] have produced guidelines to promote greater consideration of environmental, social and economic aspects in the sustainability assessment of hydro projects and the management and operation of existing power schemes. While there is disagreement on some aspects relating to WCD's detailed recommendations, there is clear acceptance of the core values listed in the report, which are equity, efficiency, participatory decision-making, sustainability, and accountability [27].

Also the values of precautionary approach and eco-efficiency to environmental management have been supported in developing sustainable hydropower. The precautionary principle and the principle that preventive action should be taken can guide hydropower decision-making and planning in the face of risk and uncertainty of the total impact of the hydropower development (see more [22, 42, 46]). In the case of the precautionary principle reference is made to the Rio declaration on Environment and Development (Rio Declaration, 1992), where it is stated that: Where there are threats of serious or irreversible environmental damage, the lack of full scientific certainty shall not be used as a reason for postponing cost effective measures to prevent environmental degradation. This statement may be extended to any risk, such an economic, ecological or social risks [16]. IHA [27] believes evaluation of power generation options should be based, where feasible, on life-cycle analysis of alternative technologies with a precautionary approach to scientific uncertainties.

Eco-efficiency is founded on the idea – doing more with less – is at the core of the business case for sustainable development. Combining environmental and economic operational excellence to deliver goods and services with lower external impacts and higher quality-of-life benefits is a key sustainable development strategy for business (see more [28, 49, 50]).

At the domestic level, good governance with sound environmental, social and economic policies, together with democratic institutions responsive to the needs of people, the rule of law, anticorruption measures, gender equity and an enabling environment for investment are the basis for sustainable development [50]. Sustainability is based on due considerations given to interrelationships and integration of competing needs. Therefore, it is of prime importance that the national and/or regional policy context takes into account cross-sectoral issues, for example through integrated water resources management (IWRM) [50].

Hydropower development must adopt the dual perspective of integrated water resources management and energy development that takes into account the broad range of social, economic, and environmental issues [47]. IWRM is about strengthening frameworks for water governance to foster good decision-making in response to changing needs and situations. Here, the rather universal definition of IWRM "as a process that promotes the coordinated development and management of water, land and related resources in order to maximize the resultant economic and social welfare in an equitable manner without compromising the sustainability of vital ecosystems" defined by GWP [18] is used (see also [8, 26, 42, 46]). Such decision support systems as IWRM appear particularly appropriate and useful in hydropower development when implemented within a participatory, transparent multi-stakeholder approach [16, 46].

3 Sustainability in the hydropower sector in Lao PDR

3.1 Environmental and social aspects of sustainability

It is important to ensure that hydropower potential in Lao PDR will be sustainably developed, which implies achieving minimal adverse social and environmental impacts, while remaining a viable, profitable and source of renewable energy supporting the country's economic development. The selection, design, construction and operation of hydropower projects need to take into account the environmental and social impacts in a manner sympathetic to the environment and society. The Government is committed to reducing dependence on fuel wood and imported fossil fuels by promoting, where practicable, renewable forms of electrical energy, especially the hydropower projects [44]. As rural Lao remains an essentially agrarian society, and the livelihoods of its people are underpinned by the presence of the healthy and diverse ecosystems that provide them with sustenance, the issue of sustainability is an important one [5].

The Government of Lao has already developed quite a comprehensive body of domestic legislation; the Law on Water and Water Resources (1996), Forestry Law (1996), Land Law (1997), Agriculture Law (1998), and Environmental Protection Law (1999) to deal with sustainability and environmental conservation issues in its policies and regulations to ensure sustainable development [48].

According to the Environmental Protection Law of 1999, and Environmental Assessment Regulations 1770 of 2000, all large hydropower projects must produce a full Environmental Impact Assessment (EIA) report and Environmental Management Plan (EMP) [39]. The Electricity Law (1997) stipulates that investors in electricity production have the obligation to protect the environment, namely to assess the impact on natural environment, on the ecosystem, to limit the impact on society and wildlife habitat [44]. Moreover, recently enacted Lao Technical Electric Standard provides all the necessaries guidance to maximize dam safety during construction and operation [24]. The Lao EIA process is largely compatible with international guidelines for conducting EIAs and the bottom line is that construction activities cannot commence until the Water Resources and Environment Agency (WREA) approval is received [10]. The social and environmental safeguards of the Multilateral Development Banks have enhanced the implementation processes of many hydropower developmental projects in Lao PDR, which lack human, technical and financial resources.

With regard to the coming Nam Theun 2, a lot has been learnt from previous experiences and issues like environmental impact and resettlement have been thought over and revised in project preparation [4]. One important policy to arise from the project has been the GOL's National Policy: Environmental and Social Sustainability of the Hydropower Sector in Lao PDR. The aim of this policy is to help ensure that the principles of social and ecological sustainability are integrated into all large hydropower developments. [39]. The establishment of the first Watershed Management Protection Authority in Lao PDR is also an accomplishment of the Nam Theun 2. This Authority is responsible for the management and protection of the watershed of the Nam Theun 2 above the reservoir formed by the Nam Theun 2 hydro-electric dam [34]. The Environmental Protection Fund (EPF) has now been established, with a specialized funding 'windows' for the collection, management and distribution of funds from large scale water developers such as hydropower [13]. However, the fund has not been functioning properly [41]. Additionally, an integrated approach to river basin management will be practiced for multiple projects planned to dam a single river [7].

Major gaps remain, however, between the formulation and implementation of legal instruments, and between the establishment and enforcement of rules and regulations. Capacity for implementation is still lacking. EIA still frequently fails to influence decision-making, like for example the case of Theun-Hinboun dam shows [5, 10, 27, 46]. Additionally, inadequate transparency and poor accountability compromise the ability to monitor the environment [6, 10]. Despite rather wide range of IWRM capacity building steps, much more remains to be done. The legal and financial basis of IWRM is still rather limited with lack of coordination and efficiency [7].

Currently there is no national water resource information management system and water related data is fragmented and scattered among agencies. Access to data and other issues are major constraints to the development of the water sector policies and strategies, as well as water resource project and to integrated water resources management [7, 45].

Public hearings in hydropower development are an essential safeguard mechanism for a meaningful intervention by consumers and other vulnerable stakeholders in decisions that will ultimately become their lasting economic and social burden [19, 46]. Citizens need to have right to question and oppose new dam construction, and have their concerns addressed in all phases of planning, construction and operation. Critics are concerned with lack of public participation in Lao PDR and that corruption and lack of democracy in Mekong countries will be increased rather than decreased by revenue streams from large hydropower projects [19].

When planning and developing hydropower sector investment, on-going development of the legal, institutional and regulatory environment and strengthening of the institutional capacity and improvements in the commercial position of Electricité du Laos (EDL) are issues to be considered in Lao PDR [15, 29]. Applying environmental considerations in the early stages of hydropower sector planning is necessary for improving the economic benefits of hydropower development and reducing subsequent project preparation time [8]. Applying sustainability considerations promotes a recognition of the strong linkage between environmental and social impacts of hydropower projects. Identifying potential social impacts at preliminary project scoping stages will facilitate consideration of alternative design options and viable livelihood development programs through a more participatory approach by avoiding the time pressures associated with the latter stages of project appraisal process and financial closure [8].

3.2 Economic aspects of sustainability

There can be no sustainable development without the demonstration of sound and equitable distribution of economic benefits. For this reason economic considerations are a central in the decision-making processes associated with hydropower projects. The efficient use of economic resources requires that the best options are selected, that alternatives have been carefully evaluated, and that there are no hidden and unforeseen costs that could emerge in the future [38].

Demand growth has been rapid and the availability of concessional funds and grants is not keeping pace with the increasing capital requirements of the sector. Recently, restructuring and reform of the energy and water sector in many countries including Lao PDR has changed the role of government in decision-making and planning, with private investors and corporations taking both financing and ownership roles in these projects [46]. Moreover, the improvement in EDL's financial position is encouraging private investors and lenders to consider national supply as well as export markets [29]. The capital and resource requirements of large export power projects require private sector participation which the Government mobilizes using Build-Operate-Transfer (BOT) modalities [44].

The financing vacuum left by the withdrawal of the development agencies has been filled for the moment by non-traditional sources of finance, notably the China Exim Bank in Lao PDR. Finance from these sources appears to be abundant and the financing model is effective in accessing loans, but it is new and the associated procurement practices are weak. [29]. NGOs and IFIs alike have raised concerns that Chinese hydropower financiers and developers generally lack policies on environmental and social issues and do not necessarily adhere to internationally accepted standards and guidelines [37, 46]. Moreover, it is difficult for policy-makers to keep pace with the scale of investment and economic growth in Lao PDR.

There are many other weaknesses in private financing models which may impact negatively to economic sustainability and effectiveness in Lao PDR. Promotion of Independent Power Producer (IPP) projects in Lao PDR begins with an unsolicited proposal from a sponsor and, from this, an MOU is drawn up and a concession ultimately negotiated. Concessions are awarded in the absence of competition after the sponsor has completed technical and environmental studies of the proposed project. [29]. Moreover, the involvement of the private sector on a significant scale introduces problems for power planners in managing the uncertainty of IPP commercial operation date (COD) and a means of countering this risk must be found for the private sector's role in domestic generation development to be dependable and constructive [29]. Delays in CODs lead to increases in interest accumulated on funds borrowed for construction activities and to delays in revenues accruing to the owner from the completed project [46].

Overall, the failure of project delivery (IPP) can be seen as one of problems related to the practice of awarding mandates for IPP projects as an unsolicited, negotiated transaction based on a BOT modality. Problems related to this approach include the lack of transparency and competition, the failure to filter out projects inconsistent with IPP program objectives, a high degree of uncertainty on project outcomes, insufficient government control of project development, and unnecessary time commitments for all parties in Lao PDR [29, 51].

When it comes to future plans for hydropower projects in the region and in Lao PDR, they are justified in part by uncertain projections of high demand for electricity in Thailand, Vietnam and China [19]. This has implicated overstating export prospects in Lao PDR. It has also been criticized that the future power targets of the EDL's Power Development Plan (2007-16) are not achievable due to lack of availability of funds for capital works and, secondly, capacity constraints within the Lao power transmission and distribution system [29].

Overstating future demand has led to a perceived need for a large incremental response to meet rapidly growing needs. In many circumstances this has militated against a gradual approach of adopting smaller, non-structural options and has pushed decision-makers into adopting large-scale dam projects because they seem to be the only adequate response to the large gap between existing supply and forecast demand. Of principal concern is that it is frequently the agencies (like in Thailand) that are responsible for building supply infrastructure that are also charged with undertaking demand forecasts, leading to a potential conflict of interest [14, 19, 46].

The opening of the sector to new financing models is fundamentally changing the way in which projects are planned and implemented. Developing new financial models require strong independent regulation and integrated resource planning [19]. There is a need to establish a regulator to set domestic retail tariffs and negotiate wholesale export tariffs. This would mean tariffs could be pre-set before bidding power generation concessions and bidding would therefore be on some other criterion, perhaps the highest royalty payments. Also, a creation of a centralized Lao power purchasing agency that could competitively bid power concessions within Lao PDR and could sell the off-take from some or all Lao projects to domestic and/or foreign power purchasers is needed [29].

Increase of the efficiency of energy use and the use of renewable sources is a priority for achieving a sustainable and equitable energy sector [46]. Lao PDR has yet very little experience implementing energy savings programs. Subsidies should be removed and, and hydropower should be forced to compete on a fair, least economic cost basis with a broad range of cleaner and less expensive alternatives [19].

4 Concluding remarks

The country's abundant water resources and mountainous terrain have allowed the Government of Lao to set up a master plan to develop hydropower and export large quantities of hydroelectric energy. The opportunities and challenges of hydropower development are complex, and ultimately dependent on the resources, skills, and will to invest responsibly, with due regard to economic, environmental and social aspects of sustainable development.

Emerging trends, driven by more sophisticated energy markets, volatile energy prices, climate change, and increased attention to water management and regional integration, are changing the value proposition of hydropower in development [47]. New financing models with increasing resources from the private sector requires a broad range of responses: better policies and institutions along with environmental and social governance (safeguards) improvement; improving payments from energy consumers; clarity in regulations for developing and operating hydro plants; and innovative financial structures that support public-private partnerships projects with multiple (public and private) benefits [47] along with improvement of negotiation capacity with hydropower developers. True least-cost economic planning and realistic demand forecast practices must become common practice among power sector planners [10].

To achieve sustainable development for future hydropower development strengthening of integrated water resources planning and management in river basins is needed in Lao PDR. Issues of balancing goals (social, economic and environmental) and of transparency and participation in planning are also important. Public participation will be a challenge for Lao PDR. There is need for more effective consultation and participation of other stakeholders including local communities in the hydropower planning system to take more effectively social and environmental issues into account.

Lao PDR can draw on the lessons that are emerging from the Nam Theun 2 project. The preparation of it has paved the way for more participatory, transparent and improved hydropower developments. These lessons can be evaluated and replicated in future projects so the best social and environmental programs are put in place in order to effectively manage impacts. The role of Mekong River Commission (MRC) as an institution that can liaise and coordinate between the varying interests of all the countries is vital in sustainable hydropower development in the Mekong basin and it needs to be strengthened. By strengthening of the role of the MRC it can more vigorously exercise its role of helping Mekong countries cooperate and promote sustainable development of its water as it takes its Initiative on Sustainable Hydropower (ISH) [32]. Also, recently the Asian Development Bank, MRC and World Wide Fund for Nature started formulating a joint project on Environmental Considerations for Sustainable Hydropower Development (ECSHD) which aims at developing a sustainability assessment tool for hydropower development that can be integrated into existing planning procedures and processes and move the input of sustainability considerations to earlier stages of the development cycle [8].

References

[1] ADB, The Economics of Climate Change in Southeast Asia: A Regional Review, Asian Development Bank: Manila, 2009.

[2] ADB, Understanding and Responding to Climate Change in Developing Asia, Asian Development Bank: Manila, 2009.

[3] ADB, Asian Water Development Outlook 2007, Asian Development Bank and Asia-Pacific Water Forum: Manila and Tokyo, 2007.

[4] ADB, Reality check of the WCD guidelines. A case study for the Nam Theun 2 hydro-electric project in Lao PDR. Workshop to Discuss the World Commission on Dams Report Dams and Development, Asian Development Bank: Manila, 2001.

[5] ADB, Environments in Transition. Cambodia, Lao PDR, Thailand, Viet Nam, Asian Development Bank: Manila, pp. 4-5, 2000.

[6] Bestari, N., Mongcopa, C., Samson, J. & Ward, K., Lao PDR: Governance Issues in Agriculture and Natural Resources. A Case Study from the 2005 Sector Assistance Program Evaluation for the Agriculture and Natural Resources Sector in the Lao People's Democratic Republic, Asian Development Bank: Manila, pp. 8, 2006.

[7] Birch, A., Benchmarking National Water Sector Capacity in the Lower Mekong Basin Countries, Discussion Paper, World Bank-ADB Mekong Water Resources Review, pp. 10-11, 2005.

[8] Bird, J., Goichot, M., Makin, I., Moua, K. & Perera, P., Environmental consideration for sustainable hydropower development in the Mekong region – a joint ADB, MRC and WWF initiative. Yangtze Symposium, pp. 4-6, 2008.

[9] Biswas, A., Integrated water resources management: a reassessment. A Water Forum contribution. Water International, 29(2), pp. 248-256, 2004.

[10] Callander, T., Environmental Impacts of Trade Liberalization in the Hydropower, Mining and Construction Materials Sectors, Lao PDR, Environment Assessment Project (RTEA) – Background Research Paper, International Institute for Sustainable Development (IISD), pp. 1-11, 2007.

[11] Department of Energy Promotion and Development (EPD). Powering Progress. Web Site, Vientiane, www.poweringprogress.org/

[12] Electricité du Laos (EDL), Draft Power Development Plan (2007-16), System Planning Office, Technical Development Committee, Ministry of Energy and Mines, Vientiane, pp. 26-27, 2008.

[13] Environment Protection Fund, www.laoepf.org.la/

[14] Foran, T. & Greacen, C., Towards More Sustainable Energy Futures for the Mekong Region: Policy Options for Development Donors, pp. 3-5, 2007, Online. www.palangthai.org/docs/TowardsMoreSustainableEnergyFuturesforMekong.pdf

[15] FREPLA2020 Project. Summary of expert workshop. Future Resource and Economy Policies in Lao PDR till 2020 First Expert Workshop, Vientiane.

The Ministry of Energy and Mines (MEM), The Finland Future Research Centre (FFRC), 2009.

[16] Ganoulis, J., Risk Analysis of Water Pollution, Wiley-VCH Verlag GmbH & Co. KGaA: Weinheim, pp. 40-200, 2009.

[17] Gerbens-Leenes, W., The hidden water consumption; virtual water. HENVI Science Day 2009: Water Use and Climate Change, 2009, Helsinki University.

[18] Global Water Partnership (GWP), Catalyzing Change: A Handbook for Developing Integrated Water Resources Management (IWRM) and Water Efficiency Strategies, pp. 7, 2005.

[19] Greacen, C. & Palettu, A., Electricity sector planning and hydropower. Democratizing Water Governance in the Mekong Region, eds. L. Lebel, J. Dore, R. Daniel & Y.S. Koma, Mekong Press: Bangkok, pp. 94-123, 2007.

[20] Gürbütz, A., The role of hydropower in sustainable development. European Water, 13(14), pp. 64, 2006.

[21] Harrison, G. & Whittington, H., Impact of climatic change on hydropower investment. Hydropower in the New Millenium, eds. B. Hanningsvåg, G. Midttomme, K. Repp, K. Vaskinn & T. Westeren, Swets & Zweitinger, Lisse, pp. 257-260, 2001.

[22] Harremoës, P., Gee, D., MacGarvin, M., Stirling, A., Keys, J., Wynne, B. & Guedes Vaz, S., Late Lessons from Early Warnings: The Precautionary Principle 1896–2000, Environmental issue report No 22, European Environment Agency, Luxembourg: Office for Official Publications of the European Communities, EEA: Copenhagen, 2001.

[23] Hydropower and Climate Change; World Commission on Dams, WCD Press Releases & Announcements, Online. www.dams.org/news_events /press333.htm

[24] Hydropower News; Department of Energy Promotion and Development (EPD), Powering Progress, Online. www.poweringprogress.org/index.php? option=com_content&view=article&id=196:22-february-2008-investors-urged-to-comply-with-lao-hydropower-standards&catid=86:hydropower-in-the-media&Itemid=50

[25] International Hydropower Association (IHA), Sustainability Guidelines, 2004, Online.www.hydropower.org/downloads/IHA%20Sustainability%20 Guidelines_Feb04.pdf

[26] International Hydropower Association (IHA), The Role of Hydropower in Sustainable Development, 2003, Online. www.hydropower.org/downloads /RoleOfHydropowerInSustDev_IHA%20White%20Paper.pdf

[27] Jusi, S., Asian Development Bank and the case study of Theun-Hinboun hydropower project in Lao PDR. Water Policy, 8(5), pp. 371-394, 2006.

[28] Lehni, M., Eco-Efficiency; Creating More Value with Less Impact, World Business Council for Sustainable Development: Geneva, 2000.

[29] Maunsell Limited & Lahmeyer GmbH, Power System Development Plan for Lao PDR (PSDP), Final Report, Volume A: Main Report, New Zealand, pp. 1-98, 2004.

[30] McCully, P. & Wong, S., Powering a sustainable future: the role of large hydropower in sustainable development. UN Symposium on Hydropower and Sustainable Development, pp. 1, 2004, Online. www.internationalrivers.org/files/irnbei.pdf
[31] Mekong River Commission (MRC), Mekong River Commission Annual Report, Lao People's Democratic Republic, Vientiane, pp. 29, 2007.
[32] Mekong River Commission (MRC). The MRC Initiative on Sustainable Hydropower (ISH) Web Site, www.mrcmekong.org/ISH/ISH.htm
[33] Messerli, P., Heinimann, A., Epprecht, M., Phonesaly, S., Thiraka, C. & Minot, N., (eds). 2008: Socio-Economic Atlas of the Lao PDR – an Analysis based on the 2005 Population and Housing Census. Swiss National Center of Competence in Research (NCCR) North-South, University of Bern, Bern and Vientiane: Geographica Bernensia, pp. 124, 2008.
[34] Nam Theun 2 WMPA homepage, www.nt2wmpa.gov.la/
[35] National Growth and Poverty Eradication Strategy

Section 5
Social issues and environmental policies

Sustainability actions in Mediterranean countries through cooperation partnerships: the case of the project PAMLED

T. Daddi[1], F. Farro[2], S. Vaglio[3], G. Bartoli[1] & F. Iraldo[1,4]
[1]*Sant'Anna School of Advanced Studies, Italy*
[2]*Department for the Relations with the European Union, Municipality of Prato, Italy*
[3]*Laboratory of Anthropology and Ethnology, University of Florence, Italy*
[4]*Institute for Environmental and Energy Policy and Economics, Bocconi University, Italy*

Abstract

In recent years the involvement of the Third Mediterranean Countries in achieving the environmental policy objectives set up by the European Union have become more and more important. There have been several programs of cooperation with co-fund activities and actions to improve the state of the environment of Third Mediterranean partners in order to achieve a global improvement of the environment. This paper aims to present the results of the project PAMLED, which was co-funded by the Med-Pact Programme of the EU. The project will complete its course at the end of April and it aims to develop and strengthen the capabilities of three Mediterranean cities (City of Marrakech, Morocco, Sin El Fil, Lebanon and Bodrum, Turkey) in managing and promoting their local sustainable development, as well as implementing innovative different action fields. The strengthening of the capabilities of these Mediterranean partners was mainly based on 'collective learning', achieved by the constitution of partnerships with five European partners (the municipalities of Prato, Lucca, Brtonigla, Rio Marina and Skopje). The needs and priorities of each Mediterranean partner were identified and pilot actions were specifically elaborated in order to promote the sustainable development and the exploitation of local resources, with particular respect to environmental protection, the enhancement of local tangible and intangible assets, economic support and

WIT Transactions on Ecology and the Environment, Vol 131, © 2010 WIT Press
www.witpress.com, ISSN 1743-3541 (on-line)
doi:10.2495/EEIA100181

overall sustainable development. The paper will show the outputs of several pilot actions carried out in the three Mediterranean cities involved. The Municipality of Bodrum carried out innovative pilot actions in the field of urban waste management; Sin El Fil developed a pilot project titled "Youth development plan", while the City of Marrakech carried out a pilot project aimed at sensitizing the local communities and the actors of the touristic sector (e.g. hotels, hammams) on the importance of reducing water consumption.
Keywords: sustainable development, Mediterranean Third countries, European Neighbourhood Policy, local development cooperation, Project PAMLED, Med-Pact Programme.

1 Introduction

The Mediterranean is the largest European sea, shared by 460 million people living in its 22 countries and territories and it is visited by some 275 million more every year. According to the 2009 UNEP/Plan Bleu [1], the shores of the Mediterranean basin account for 5.7% of the world's land mass, 7% of the world's population and 12% of the world's GDP. However, despite its richness, the Mediterranean region represents one of the most vulnerable environments in the world, also accounting for 60% of the world's "water-poor" population and 8% of global carbon dioxide emissions. Sustainable development is a global objective for the Mediterranean countries.

The Mediterranean coastal regions are also the part of the basin, which has undergone the most dramatic alterations due to the rapid change of demographic trends, the new socio-economic conditions prevailing (which favour higher consumption of natural resources), and also due to new technologies, including transport (new roads, new types of ships, new harbours, etc) [2]. In addition, an increasing waste production and pollution, closely related to a conspicuous loss in biodiversity, was evidenced [3]. A greater commitment in the reduction of inequalities and also in assisting the poor countries' development was emphasized and expected at a global level [4]. Moreover, the need to change unsustainable production and consumption patterns was warmly advised in order to protect and manage natural resources sustainably, to safeguard health and integrate the objective of sustainable development more effectively into the process of globalization, as confirmed by the Marrakech process [5, 6].

The Mediterranean countries, mostly the regions of the South and East, can plan and manage their development in a sustainable perspective. The potential increase in environmental pressures on Mediterranean coastal regions over the coming 20 years is considerable, particularly in the tourist sector: Indeed, tourism has induced an increase not only in transport, but also in urban development and sprawl, energy infrastructures, etc [7]. In addition, a continued spread of unsustainable production and consumption patterns is likely to increase the environmental costs dramatically, which already account for 3-5% of the GDP [8]. The environment should not be considered as an additional constraint, but as a driving force, an asset and an incentive to improve local development [9].

The Mediterranean peoples are now much more aware of the threats to their environment and their unique natural and cultural heritage, as evidenced by the high number of policies that have been adopted in almost all Mediterranean countries in order to provide specific solutions to these problems [10]. Most Mediterranean economies, which have been insufficiently dynamic in the last 20-30 years compared to other regions of the world, today are experiencing critical level of unemployment (ranging between 8% and 25%). Social issues are also a major concern, particularly in the southern and eastern Mediterranean countries: in spite of progress, they are still backward in terms of literacy and gender equality. The pattern of economic growth of the Mediterranean Partner Countries is increasingly reliant on the ability of their industrial activities to face up to the competitive challenges of the EU markets, by complying with increasingly high quality standards and performance requirements. On the one hand, this is curbing a phenomenon known as the Kuznets curve, which explains how the increase in production and economic output of a country can lead to a corresponding increase in the polluting emissions and resource consumption, particularly in the early phases of the industrial development process. Therefore, the industrial growth is deemed to negatively impact on the environment. On the other hand, in order to be fully integrated in the economy of the Developed Countries and have access to the EU market at socially acceptable conditions, the industrial production of the Mediterranean countries, and the products they offer, must increasingly comply not only with performance and quality standards, but also with environmental quality requirements.

The challenge of globalization requires widespread regional cooperation, political stability, efficient governance and social protection. Yet the situation of the Mediterranean countries in fulfilling these conditions is very asymmetrical. The Mediterranean European Countries are facing the challenges of globalization with the strong backing of the EU. The southern and eastern Mediterranean countries, which are of course faced with the same challenges of globalization, do not benefit from such dynamic regional cooperation. Established in 1995, the Euro-Mediterranean Partnership still needs a collective vision of sustainable development, besides appropriate resources and commitment [11]. This situation is closely related both to the inadequate levels of North-South and South-South cooperation and also to the continuing conflicts, especially in the Near East, even though some longer-term political solutions appear to be emerging. If the relevant reforms will not be quickly implemented, the differences between the two shores of the Mediterranean will result in growing instability and may accentuate the existing levels of social and economic asymmetry. A possible alternative is to maximize complementarities and opportunities between the North and the South, in the context of joint and differentiated processes of sustainable development, to optimize the positive effects of globalization. The European Neighbourhood Policy (ENP) could be the right direction [12–14]. The ENP seeks to deepen political cooperation and economic integration between the EU and its immediate neighbours and to promote and support better governance and reform in Mediterranean countries. Through mutually agreed Action Plans and cooperation projects, the EU and its ENP partners will address issues of common interest and devise measures beneficial to economic growth and social cohesion, raising living standards and protecting the environment, thereby contributing to the long-term goal of sustainable development in the Mediterranean region [15].

2 Project PAMLED – background

The Mediterranean basin posses a wide range of different and frequently critical socio-political, economical and environmental assets, thus strengthening cooperation among the Mediterranean countries to support their development, offers an effective tool to promote and enhance their natural, cultural, economic promotion and local sustainable development pursue. In this general framework the Project PAMLED ("Building Effective Partnerships among European and Mediterranean Municipalities for Local Economic Development Promotion") was conceived in 2006 and carried out since.

The Project PAMLED is co-financed by the European Commission within the funding lines of the Med-Pact Programme ("Local Authorities Partnership Programme in the Mediterranean") in an amount of €450,000 to promote the cooperation among Mediterranean countries (overall amount €562,500 of all partners co-financing).

Taking into account the peculiarities of the different partners involved, the Project PAMLED aimed to develop and strengthen the capability of three Mediterranean partners (the City of Marrakech, Morocco, and of the Municipalities of Sin El Fil, Lebanon and Bodrum, Turkey) to manage and promote their local development, and also to implement innovative ways to respond to their problems and issues in different action fields (i.e. "Planning and Management" and "City Marketing"). The strengthening of the Mediterranean partners' capabilities was achieved by applying a principle of 'collective learning', by establishing lasting partnerships with five European partners (Municipalities of Prato, Lucca, Brtonigla, Rio Marina and the City of Skopje), as well as exchanging the most significant experiences and expertises gained in the two abovementioned macro sectors by the European partners. Needs and priorities of each Mediterranean partner were previously identified, and specific pilot actions elaborated in order to contribute to a sustainable development and exploitation of local resources. In this context, the proactive involvement of the civil society played an important role to implement the pilot projects, contributing to bring in added value to the whole PAMLED. Particular care regarded the promotion of environmental protection, the enhancement of local tangible (environment, cultural heritage, planning and promoting of economic potentialities, etc) and intangible (culture, traditions, etc) assets.

In the following sections we will detail the different pilot projects carried out by the three Mediterranean partners, and the main results achieved in different fields of action.

In designing the pilot initiatives a central issue attained the bridging of the gap between theory and practice while focusing on priority issues for the Mediterranean partners. This approach intended to avoid an inefficient overlapping with existing national or international interventions and to integrate them with other actions already being implemented in each of the three territories.

In fact, a process of validation of the pilot projects as proposed by the Mediterranean partners was necessarily carried out. It involved the representatives of the local authorities, through a participatory approach with all other potential stakeholders and actors of each territory, and the implementation of visits of technical assistance by the PAMLED teams of the European Partners.

2.1 Youth development strategy in Sin El Fil (Lebanon)

In Lebanon, as a consequence of the Taif agreements of 1989, that represented the end of the civil war after more than 20 years, an administrative reform to assist a process of decentralization of administrative competencies between the central and municipal levels of government has been envisaged among the necessary steps to support the return to normalization of the Country.

The administrative reform focused on the importance of decentralization to allow a wider participation and empower local authorities to manage their territories and reach the citizens compared to the policies implemented by the central government level. Unfortunately, this reform has not entered into force yet despite the various laws and the pressures by national and international stakeholders. It means that the typical functions often granted to the local authorities (municipalities are around 950 in a country of less than 4 million inhabitants) are often managed by the central government, thus creating a dysfunction in the distribution and management of financial and economic resources, an overlap of competences among the various levels of government that tends to remove the responsibility and legitimacy of local administrators on their political-technical decisions. Besides, this approach tends also to cause a sense of detachment of citizens from their own Municipality that is not considered as a central reference for public services, but rather a mere "tax-collector", and a bestower of incomprehensible bureaucratic procedures.

As highlighted also by the OMSAR [16], the lack of professionals and skilled civil servants is also one of the main obstacles to the application of the administrative reform in a country that highly demands decentralization to meet the needs of all citizens despite their religious belonging.

In this framework, the Municipality of Sin el Fil (metropolitan area of Beirut), represents a good exception to this scenario. The actions promoted by its pilot project demonstrated that it is possible to start-up administrative and strategic reforms to work in a more pro-active way for the wellbeing of the community.

When stressing its priorities within the PAMLED, the Municipality of Sin El Fil focused to review its economic potentiality and planning, while working strategically to support the growth of specific economic categories.

Initially, the Municipality of Sin El Fil intended to propose an ambitious pilot project aimed at setting up a Youth Business Incubator. In fact, due to the length of the administrative procedures needed to identify the suitable building and allocating the necessary internal human resources to manage the initiative, the Municipality, in consultancy with its European Partners and technical experts, agreed to re-define its original plan to respond to the economic needs of its young population, by planning a mid-long-term urban and social strategy, instead.

Therefore, Sin El Fil defined its Pilot Project in the field of economic development and strategic planning by focusing on youth entrepreneurship as an approach to reduce the current flow of brain-drain affecting its territory. The plan is based on a similar approach experienced by the Municipalities of Prato, Skopje and Brtonigla, where similar "Youth Action Plans" have been experienced in Croatia and Skopje in the years 2000, and in Prato since 2006, taking into consideration the different socio-political and historical backgrounds of the these areas.

To achieve its goal, the Municipality of Sin El Fil required a multi-direction strategy targeting potential young entrepreneurs drafted in an *ad hoc* business opportunity plan comprising three fields of action: i. Social Economic Analysis in Sin El Fil Baldeh (one of the four suburbs of Sin El Fil); ii. Local Development Youth Plan; iii. Youth in Business (Open Competition on the Best Business Ideas).

Applying the same city-to-city partnership approach envisaged by PAMLED, the technical staff of the Municipality of Prato and of the City of Skopje accompanied Sin El Fil to start-up its pilot project that outlined a youth entrepreneurship strategy as an approach to reduce the brain drain in the area; as well as identifying the bases of a future comprehensive strategic plan focusing on their socio-economic development.

The city-to-city approach offered good practices on models that provide resources, support and opportunities in the areas of employment and education, helping youth identifying their skills, preparing for job interviews and entering the job market. Great emphasis was given to the importance of communicating to youth and to spread the voice about the competition at local level. This was a crucial element with meetings being organized among the municipality, universities and high schools to motivate students to participate. It represented an important action to be carried out in a context where usually there is a huge need for bridging local communities with their administrators.

The direct beneficiaries targeted by the Pilot Project included: 2,800 households interviewed and monitored; 1,200 business stakeholders interviewed and monitored; around 50-60 young people involved; citizens' associations, local economic stakeholders involved; the University of Lebanon; the Municipality of Sin El Fil internal staff (18 councillors, 40 administrative/technical staff); the municipal staff of the Cultural Centre (around 10 staff members).

Specifically, the activities carried out envisaged: *Action 1 - On-field data collection to assess Sin El Fil Baldeh social-economic situation* - drafting of questionnaires and software development to analyse the data collected, selection and training of the interviewers, a communication plan addressed to the community; *Action 2 - Development of the Local Development Youth Plan* - an analysis of the social-economic outputs, the identification of specific actions on entrepreneurship, community meetings to facilitate the participation; *Action 3 - Youth in Business* - to appoint a Supervising Tutor, map of the business sectors in Sin El Fil, establish a technical Coordination Board to support the Municipality in managing and monitoring the action; opening of a public Call for Proposal on Business Competition for Young Entrepreneurs (providing ad hoc training of the applicants, selection of the Best Business Ideas and an initial financial support of 1,500 US Dollars to start-up their businesses).

The main outputs, strengths and weaknesses can be summarised as follows: 1) Sin El Fil ensured a strong level of participation and capability to involve and mobilize the territory and stakeholders (youth/local entrepreneurs, etc), demonstrating an in-depth knowledge of the priorities and needs of the population, far beyond the responsibilities actually assigned to local authorities in Lebanon. This approach convinced Sin El Fil to co-finance the project from the Municipal budget, even if it was not required by the project itself.

Moreover, great efforts were made to "bridge" the local community with the local administration and attention was paid to ensure that the community involved benefited by the project, although with great difficulty since citizens do not expect this kind of approach from their Municipality.

A positive outcome resulted also in the high involvement of the local technical stakeholders (local enterprises, local business men, local universities, etc) who had not been targeted initially, but accepted to become part of it.

Meanwhile, the socio-economic mapping of the suburb of Baldeh in Sin El Fil offered a good response with 60% of households answering the questionnaires drafted by the Municipality in co-operation with sociologists and social workers, despite people's scepticism towards the Municipal officers, in a country where the last population census applied in 1932.

A weak element has been noticed in the poor involvement and awareness of the central government towards the efforts made by its municipalities. In future, specific actions to attract the interest of the central government levels should be envisaged to reinforce the institutional aspects and guarantee sustainability and eventually the possibility to replicate this kind of project in other parts of the country. However, there are positive elements to consider the Project's multiplier effects and sustainability. In fact, the municipal officers developed a solid capacity to manage their economic potentiality and care about the social needs of the community in strategic terms, planning for the medium/long-term, rather than coping with the management of the municipal problems through "spot actions". They actors involved experienced new skills in project management and in approaching their local community; coordinating their efforts internally, and promoting them externally in the territory. The *motto* became "we, the Municipality, are here for you".

2.2 Development plan of environmental communication to sensitize citizens on water consumption in Marrakech (Morocco)

The Urban District of Marrakech, extended over an area of 190.42 km^2 and with a population of 877,500 inhabitants in 2003, is a decentralized territorial entity with legal status and financial autonomy. The City of Marrakech has been managed by a single Municipality (Municipal Charter of 1976). Marrakech is located in the region of Marrakech-Tensift-Al Haouz, one of the 16 regions of

Morocco. Before 1992 the urban district of Marrakech was unified (Municipality of Marrakech), while between 1992 and 2002 the City of Marrakech was composed of five urban districts: the Municipality of Guéliz Menara, the Municipality of Medina, the Municipality of Méchouar Kashba, the Municipality of Ennakhil, the Municipality of Sidi Youssef Ben Ali and of another urban district.

In June 2007 the City of Marrakech achieved the environmental certification ISO14001:04, as the first African city and of the Arab world to have implemented an EMS certificate by an accredited third party organization, through an international project funded by the European Commission, within the LIFE-Third-Country Programme titled "Marrakemas".

According to the policies undertaken at the local and national level, and in light of the results obtained during the previous stages of the Project PAMLED, the pilot project, entitled "Development plan of environmental communication to sensitize citizens and carrying out of a sample of actions envisaged by the plan in the district of Guéliz of Marrakech" was drafted and implemented. The main objectives of the proposed pilot action were to contribute to the environmental education of the population for a sustainable and rational use of water (such as the decrease of water losses both in public and private spheres, the development of processes for the control of water consumption) by carrying out a public awareness campaign, and promoting eco-sustainable tourism encouraging the use of sustainable resources (particularly water). These objectives were in line with the evidenced needs to spread a new culture aimed at the respect for the environment and with a sustainable use of resources. The pilot actions were implemented in a selected district of the Marrakech city, the suburb of Guéliz, chosen on the bases of the results of an analytical study, carried out by the City administration in association with a local institute of technical expertise (Energy Concept srl) in 2007 and 2008. The pilot actions were particularly addressed to the citizens of the suburb, its main public service companies and local actors that play an important role in the most characteristic and water management impacting sectors (Riads owners, Guest houses, Moorish Baths, Schools, Handicraft Trades, the RADEEMA, Companies delegated to clean the water basin, Associations for environmental protection in the district of Guéliz). The implementation of this pilot project profited of the suitable partnership with the city of Lucca (Italy), thanks to its experience in the sustainable use of resources, and also to its traditional environmental engagement. Moreover, Lucca had already been partner with Marrakech in the city certification-process. The fact of having once again a partnership with Lucca can be viewed as a way to ensure and reinforce the continuity of the environmental process by controlling the most important environmental aspects, in this case not only on the level of the Municipality, but even at the level of the city. Thanks to the implementation of the pilot project, some interesting results have been achieved.

Firstly, a better knowledge of the use of waters and of the higher pressure sectors in terms of water consumption and the identification of specific measures and action for a sustainable management of waters. This result was obtained by distributing specific questionnaires, and by implementing specific systems to monitor and measure the water consumption. The implementation of these devices permitted to regulate the flow of undeclared wells near the water basin. The implementation of the pilot project resulted also in an increased awareness and perception of the water problem by citizens and stakeholders alike, thanks to an efficient awareness campaign with public stands, posters, door-to-door leaflets. This action resulted also in an increased mobilization and awareness of some local actors around the water problem. Long-term results will comprise also a decrease in water costs and the extension of this water management model on the whole town or on other areas.

2.3 Waste management and tourism development in Bodrum Marina (Turkey)

Thanks to the mechanism of "city-to-city partnership", developed under the Project PAMLED with the Municipality of Lucca (Italy), the Municipality of Bodrum carried out also a pilot project titled "Environmental Awareness Action Plan".

At this purpose, first of all a decision was made to determine the target area: with this objective the area between the Marina and the Castle was selected as the project area. This area was considered ultimately suitable for information and awareness campaigns since the local population use this area intensively, allowing visual and informative materials to create higher performance results. The area had been visited to quantitatively determine the number of users, the waste collection potential resulting from the waste production by its users, and the number of bins, containers, depots and collection bags available. Then, a decision was made to determine the necessary quantities and costs in cooperation with the garbage collection company to place special-design bins adhering to ease-of-use and city safety rules to collect vegetable oil waste in the target area containing numerous restaurants and daily tour boats. Collectors have been involved in decisions through a process of holding regular meetings with the technical team of the local Municipality. Although it was mandatory according to the Turkish "Regulation on the Control of Packaging Waste", due to the lack of people's habits on waste differentiation and collection at the source, it was decided to add statements about penal provisions if incurring in violations of the rules and regulations, on all brochures and fliers distributed to boats and residences and, moreover, to clarify the article of Law applicable in this case. An operation concluded by consulting with the Municipal Legal Department.

Successively the following activities were carried out: (i) informative and awareness raising efforts were carried on by using local mass media (radios, TV, newspapers); (ii) communication channels - including billboards, warning signs, bulk mobile text messages and e-mail messages - were used in addition to special information and warning brochures; (iii) special informative campaigns were carried out towards residences, restaurants and boats, progress and development were determined on site through regular visits and, moreover, campaigns were carried out in a user-friendly manner; (iv) in order to experience the good practices (approach of city-to-city partnership), the representatives and technical experts of the Municipalities of Bodrum and Lucca visited their respective towns in 2008 and 2009; (v) press conferences were held to develop awareness among the community on waste solid collection.

Consequently the results of this pilot action can be summarized as follows: (i) drafting of an awareness plan of action on solid waste, recycling and pollution among local people; (ii) establishment of a trend of active participation by local stakeholder in the implementation of pilot project activities; (iii) contribution to the acceleration of the recycled collection activities already started throughout Bodrum.

The pilot project represents a good input for the Cleaning Department which was created in 2005. Until then, in spite of some spot actions, the Department did not have previous opportunities to start up a real recycling policy on the territory. Before the present project had been implemented, only some bins were located in the city, but not a real campaign was possible to support their efficient use. Instead, the current project focused both on the dissemination of proper bins for recycling (material tool) and on the information/awareness aspects of its use. The project is placed in a context in which national authorities are demanding Municipalities to implement recycling process at local level. Therefore, it could represent a good step to give evidence of the municipality skills and to be ready to extend the same activities to other areas of the city. The municipal internal staff was directly involved in the campaigns and local people know them and ask for help when needed. More specifically, the Cleaning Department staff were directly involved in the awareness meetings: this allowed a direct contact between people and the municipal staff. A telephone number is highlighted on the campaign poster and, for instance, people call up when bins are full. Anyway, the most important element of the project regarded the increase of people's awareness. Indeed, it did not focus just on providing bins or the other material disposals but large efforts concentrated on awareness raising tools. The internal staff of the Municipality of Bodrum explained to the population involved how the recycling system was being implementing in order to facilitate its use. The campaigns were being currently implemented both along the Marina and in the schools.

On the other hand, one weak factor regards the current poor financing availability for the Municipality of Bodrum to extend the pilot project to other areas of its territory. Furthermore, a real long-term recycling process is not possible at the moment at local level because of lack of a recycling plant in the Bodrum Peninsula. Even if waste collection would be possible - at least in the area where the pilot project has been implemented - the recycling process has to be reinforced through additional and sustainable financial resources and a sound coordination at regional and national level in Turkey.

3 Conclusion

The bottom-up local development model that encourages integration and synergies at local level, as well as among private, public and civil society actors, had its main *raison d'être* and development in the identification of the content of each Pilot Project, in line with the existing priorities of each territory.

The objective of strengthening the capability of the three involved Mediterranean Municipalities in planning and managing local economic and environmental development initiatives has been achieved by setting up operative and lasting partnerships among the cities involved in the action and implementing pro-active interventions in two specific fields of action: "Planning and management" and "City marketing". Thematic city-to-city partnerships allowed Municipalities to interface and share problems with other actors operating at the same level and with similar degrees of empowerment. As n-depth institutional analyses were conducted along each Municipality, potentially resulting in the establishment of new approaches and methods to gather local needs and to design local development policies. Procedures and practices for local level resource mobilization and public expenditure management were encouraged.

Developing a new role and a pro-active involvement to address local needs generated new approaches within the internal organization of the Mediterranean partners. Mutually beneficial and collaborative relationships between the European and the Mediterranean partners were initiated through "working groups" and the "city-to-city" networks. Activities that evinced partner potentialities and expertise gained on local development management to be shared with other cities. Furthermore, the PAMLED Scientific Committee bridged the gap among the Municipalities and provided them the knowledge and skills necessary to achieve these goals. Pilot Projects were identified on a proven basis of sustainability. During the pilot actions drafting phase, particular attention was given to single out indicators demonstrating their future sustainability, while Municipalities were encouraged to make out interventions that could fit in other institutional goals or initiatives. Several activities were organized to better incorporate local citizens and stakeholders' points of view.

Some multipliers effects were pursued (and at least partially reached): dissemination of efficient, innovative and customized practices for local economic development support; positive spill-over effects along other Municipal Departments and other relevant stakeholders; progressive economic integration process; positive indirect employment and income generating effects of the additional people benefiting from outputs and services delivered by the pilot projects.

References

[1] United Nations Environmental Programme (UNEP), MAP-Plan Bleu, State of the environment and development in the Mediterranean, 2009.

[2] Scoullos, M.J., Impact of anthropogenic activities in the coastal region of the Mediterranean Sea. *International Conference on Sustainable Development of the Mediterranean and Black sea Environment*, 2003.

[3] Hall, C.M., Tourism urbanisation and global environmental changes. *Tourism and Global Environmental Change. Ecological, Social, Economic and Political Interrelationships*, eds. S. Gössling & C.M. Hall, London: Routledge, pp. 142-158, 2006.

[4] Du Plessis, C., Action for Sustainability: preparing an African plan for sustainable building and construction. *Building Research & Information*, 33 (5), pp. 1-11, 2005.
[5] Mbohwa C. & Fukada S., ISO 14001 Certification in Zimbabwe: Experiences, Problems and Prospects. *Corporate Environmental Strategy*, 9 (4), pp. 427-436, 2002.
[6] Clark, G., Evolution of the global sustainable consumption and production policy and the United Nations Environment Programme's (UNEP) supporting activities. *Journal of Cleaner Production*, 15 (6), pp. 492-498, 2007.
[7] Butler, R.W., Tourism, Environment and Sustainable Development. *Environmental Conservation*, 18 (3), pp. 201-209, 1991.
[8] United Nations Economic Commission for Europe (UNECE), Annual Report Economic Essays, 2009.
[9] Khan, Z., Cleaner production: an economical option for ISO certification in developing countries. *Journal of Cleaner Production,* 16, pp. 22-27, 2008.
[10] Ernoul, L., Residents' perception of tourist development and the environment: a study from Morocco. *International Journal of Sustainable Development and World Ecology,* 16 (4), pp. 228-233, 2009.
[11] Marks, J., High hopes and low motives: The new euro-Mediterranean partnership initiative. *Mediterranean Politics,* 1 (1), pp. 1-24, 1996.
[12] European Commission, European Neighbourhood Policy. Strategy Paper, Brussels, May 12, 2004.
[13] European Commission, Proposal for a Regulation of the European Parliament and of the Council laying down general provisions establishing a European Neighbourhood and Partnership Instrument, COM(2004) 628 final, 2004.
[14] European Commission, Communication from the Commission to the Council and the European Parliament on Strengthening the European Neighbourhood Policy, COM (2006)726 final, 2006.
[15] Browning, C.S. & Joenniemi, P., Geostrategies of the European Neighbourhood Policy. *European Journal of International Relations,* 14 (3), pp. 519-551, 2008.
[16] OMSAR Lebanon, Office of Minister of State for Administrative Reform Beirut, Annual report, 2008.

Relevance of environmental and public safety issues predicts public importance of economic vitality

R. Thomas, S. Conway, P. Washeba, R. Cameron & R. Skidmore
Urban Environmental Research, LLC, USA

Abstract

The extant literature indicates that in times of economic hardship, people afford less attention to environmental issues and, intuitively, more importance to economic matters. One topic that is not provided in the literature is the relative individual importance of environmental issues; specifically, that more personally relevant issues may remain imperative in times of economic hardship. Additionally, persons may be more concerned about basic needs, such as public safety, compared to environmental issues. In this study, respondents were 600 adults during a recently documented recession. Participants indicated the most urgent environmental issue as water concerns. Due to the distress of water availability in the American southwest, the personally relevant environmental issue was conceptualized as water conservation. In order to maintain standardization of environmental issues across relevancy; another water related issue was chosen as the personally irrelevant environmental variable. Flood control, while vitally important due to urbanization, is not personally relevant. The other predictive variable, public safety; was measured as the importance of maintaining low crime rates. The prediction of the economic vitality variable was constructed as the importance of improving job opportunities, with higher importance of improvement indicating a poorer semblance of vitality. Data was analyzed using standard multiple regression and the results support the hypothesis. Specifically, in times of economic adversity, more importance is given toward public safety and personally relevant environmental issues, while less importance is focused on environmental issues that are personally irrelevant. The results and discussion are a pragmatic look into importance and perceptual influence on economic, public safety, and environmental issues.
Keywords: environmental issues, public safety, and economic vitality.

WIT Transactions on Ecology and the Environment, Vol 131, © 2010 WIT Press
www.witpress.com, ISSN 1743-3541 (on-line)
doi:10.2495/EEIA100191

1 Introduction

Economic development

The goal in economic development is to achieve and maintain economic welfare for individuals and sustain growth (Okina [13]). The conventional economic prosperity theory posits growth should be based on capital development, population increase, and advancements in technology instead of quantitative aspects such as monetary constructs. Some research shows, however, during times of crisis the way in which economies are valued is changed. Specifically, economies tend to be focused on more basic needs such as planning and preparation during instability (Perelman [14]). Further, some research suggests that as economies develop and augment, policies correspondingly adjust. This has been demonstrated as a concomitant increase in domestic environmental policies and economic expansion (Anderson [2]). That is, as an economy develops and grows, higher-level policies, such as environmental, garner more attention and amplification as well. Correspondingly, environmental issues are of higher importance in industrialized countries, which have basic needs met, as compared to developing nations who may worry about lower-level needs such as those for food, clothing, or shelter (Dunlap et al. [3]). One of the most influential theories in motivation and needs research proposed by Maslow [10–12] suggests a hierarchy of needs motivates persons. The needs in this hierarchy in ascending order consist of physiological, safety/security, social, ego/esteem, and self-actualization. According to the model, usually persons are required to fulfil lower order needs before progressing to higher order needs. That is, people will remain in their position until lower-level needs are met, at which point they will ascend the hierarchy. The highest order needs are associated with higher-level thought and meaning. Conversely, issues relating to basic needs are lower in the hierarchy and require less cognition. This research would suggest that at varying levels of economic development and prosperity, elements such as importance, focus, and attention correspondingly vary. One question posed by this variation in economic vitality and importance is, how does relative importance to persons affect these variables? One issue of limited investigation in the literature is the relative individual importance of environmental issues.

Environmental concerns

The existing literature indicates that in times of economic privation, people afford less attention to environmental issues and, intuitively, more importance to economic matters (Dunlap et al. [3]). Research suggests that while the United States is supportive of environmental issues as a nation, environmental concerns are not as salient as other issues and, therefore, are not considered as important. This suggests that issues have the potential to be placed on a continuum, thus creating varying levels of particular constructs (i.e. levels of relevance, importance, and salience will vary). Correspondingly, environmental issues have been referred to as "luxury" concerns because they are not of basic importance (Dunlap et al. [3]). That is, persons are less likely to support environmental initiatives during tough economic times on an individual level, and some research shows that there is a difference between importance of national economy and environmental issues, with the latter being less imperative (Lee and Norris [8]; Frazen and Meyer [5]). In order to ascertain the explanation between importance and issues, some research has investigated the difference in personal relevance in relation to environmental issues.

Research examining the personal relevance of environmental issues has found that in the majority of nations surveyed, environment has obtained a fair amount of concern (Dunlap et al. [3]). Environmental issues affect quality of life and environmental preservation is a key component to current political discussions (Kakrida [9]). Issues that affect quality of life are considered to be personally relevant because they directly affect the individual. The relative personal importance of environmental issues may influence the outcome and significance of the interactions of the variables. Specifically, environmental issues that are more personally relevant may remain important in times of economic hardship, while others that are less individually relevant would be less important. As evidence, environmental issues are of higher importance in industrialized countries, which have basic needs met, as compared to developing nations which may worry about food, clothing, or shelter (Dunlap et al. [3]). For those persons who may be concerned about basic needs, public safety, a very fundamental need, would plausibly be more important than environmental issues.

Public safety concerns

In addition to economic evaluation during times of basic need preoccupation, issues that are of public safety concern are considered to be more salient and important compared to environmental issues (Dunlap et al. [3]). Further, based on varying levels of economic stability cross-nationally, some nations will focus on more basic economic needs such as unemployment instead of higher-level concerns such as environmental issues (Lee and Norris [8]). Once more, it is posited that higher-level concerns are not concentrated upon until mandatory needs, such as safety, are met. Giving credence to this postulation, post materialistic values, which are considered to be higher-level, are associated with environmental concern (Lee and Norris [8]). Frazen and Meyer [5] discuss that environmental concern is associated with post materialistic values and persons with higher incomes in wealthier nations give more consideration toward environmental trends, issues, and concerns compared to those with lower socio-demographic variables.

Research has shown that public safety remains of high importance to persons regardless of economic conditions (UNLV Cannon Survey Center, Las Vegas Sun [17]. As demonstrated by Maslow [10–12], persons are required to fulfil low order needs before progressing to high order needs. The highest order needs are associated with higher-level cognition. Pertinent to the current investigation, topics relating to environmental issues require higher-levels of discernment. Conversely, issues relating to basic needs, such as those for safety, are lower in the hierarchy and require less cognition. During times of privation more basic needs are important over and above those that are non-essential. The basic need of public safety, i.e. maintaining low crime rates, would be considered more important than nonrelevant environmental issues, but as important as those distinguished as relevant.

Hypotheses
The current paper provides an empirical research investigation into the perception of economic vitality which was hypothesized to be predicted by the importance of a personally relevant environmental issue and a public safety issue. In addition, the importance of a personally irrelevant environmental issue was hypothesized not to predict economic vitality. Previous research on this topic provides correlation data corroborating the hypotheses, and the current study utilizes experimental research design.

2 Methods

2.1 Participants

Participants were 600 adults (296 males and 304 females) recruited from an online database of landline telephone numbers during a recently documented recession. Participation was voluntary and no monetary compensation was offered. The median age of the participants ranged from 25 to 44 with an overall range from 18 to 65. The median income ranged from $40,000 to $60,000 USD.

2.2 Procedures

Participants were called and requested to answer questions regarding issues of importance to the residents of Southern Nevada in the United States of America. It was indicated that all responses would remain confidential and grouped for analysis. Upon agreement of participation, respondents answered questions regarding public safety, environmental, and economic issues. Subsequent to the completion of the survey, participants were thanked for their time and asked for demographic information including age and income.

2.3 Measures

To determine the relevant environmental issue, participants were asked to indicate the most urgent environmental issue affecting life, with the largest percentage stating water concerns (45.7%). The subsequent most common responses were overpopulation (16.3%), air quality (15.3%), and energy use (6.2%).

All measures, less the obtainment of water concern, were continuous variables with endpoints ranging from 1 (low) to 5 (high). Due to the distress of water availability in the American southwest, the personally relevant environmental issue was conceptualized as water conservation. Participants were asked "on a scale of one to five, where one means 'low importance' and five means 'high importance,' please rate the level of importance for water conservation.

In order to maintain standardization of environmental issues across relevancy; another water related issue was chosen as the personally irrelevant environmental variable. Flood control, while vitally important in the American southwest due to urbanization, is not personally relevant. Participants were asked "on a scale of one to five, where one means 'low importance' and five means 'high importance,' please rate the level of importance for flood control.

The other predictive variable, public safety; was measured as the importance of maintaining low crime rates. Participants were asked "on a scale of one to five, where one means 'low importance' and five means 'high importance,' please rate the level of importance for maintaining low crime rates.

The prediction of the economic vitality variable was constructed as the importance of improving job opportunities, with higher importance of improvement indicating a poorer semblance of vitality. Participants were asked "on a scale of one to five, where one means 'low importance' and five means 'high importance,' please rate the level of importance for improving job opportunities.

2.4 Summary

Several independent variables were created in order to predict the outcome of the dependent variable. The predictive, or independent variables, in the current study were a public safety variable, measured via maintaining low crime rates; a personally relevant environmental issue variable, measured by water concern; and a personally irrelevant environmental issue variable, measured by flood control. The outcome, or dependent variable, was economic vitality, measured by improvement of jobs. All variables were analyzed via multiple regression analyses using a statistical software program.

3 Results

The survey had a margin of error of ±4 percent at the 95 percent confidence level. The data was collected for the purpose of the Clark County Monitoring Program (monitoringprogram.com). Data was analyzed using standard multiple regression analyses and the results support the hypothesis that economic vitality is predicted by personal relevance of the environmental issues and public safety. Specifically, in times of economic adversity, more importance is given toward public safety and personally relevant environmental issues such as water conservation, while less importance is focused on environmental issues that are less personally relevant such as flood control.

3.1 Environmental issues

Relevant environmental issue. Participants were asked "on a scale of one to five, where one means 'low importance' and five means 'high importance,' please rate the level of importance for water conservation ($M = 4.11$, $SD = 1.30$). Results indicate the relevant environmental issue significantly predicts economic vitality (standardized beta = .185, $t = 4.65$, $p < 0001$). Specifically, as importance of economic vitality increases, so does the importance of the relevant environmental issue.

Irrelevant environmental issue. Participants were asked "on a scale of one to five, where one means 'low importance' and five means 'high importance,' please rate the level of importance for flood control (M = 3.70, SD = 1.21). Results indicate flood control did not significantly predict economic vitality (standardized beta = .045, t = 1.176, p = .240). Specifically, there was no relation between increases in importance of economic vitality and the irrelevant environmental issue.

3.2 Public safety issue

The other predictive variable, public safety; was measured as the importance of maintaining low crime rates. Participants were asked "on a scale of one to five, where one means 'low importance' and five means 'high importance,' please rate the level of importance for maintaining low crime rates (M = 4.14, SD = 1.11). Results indicate public safety significantly predicts economic vitality (standardized beta = .319, t = 8.159, $p < .0001$).

3.3 Overall model

Data was analyzed using standard multiple regression analyses and the results support the hypothesis that economic vitality (M = 3.92, SD = 1.30) is predicted by personal relevance of the environmental issues and public safety (R square = .176, R = .420). Specifically, in times of economic adversity, more importance is given toward public safety and personally relevant environmental issues such as water conservation, while less importance is focused on environmental issues that are not personally relevant such as flood control.

4 Discussion

It was hypothesized that a personally relevant environmental issue and a public safety issue would predict economic vitality during economic privation. In addition, it was predicted that a personally irrelevant environmental issue would not predict the economic variable. Results support the hypotheses. That is, an increase in importance of the personally relevant environmental issue was associated with an increase in importance of economic vitality; correspondingly there was no association with the irrelevant environmental issue and economic vitality. This study corroborates previous research and extends the scientific literature by addressing the hypotheses and theory through an empirical research design. One question posed by these findings is, why did type of environmental issue differ in predictive ability?

Importance of environmental issues is contingent upon relevance. Due to the higher-level cognitions involved with environmental processes, only those issues that are germane to the individual are considered important, particularly during times of economic deficiency. This research corresponds with past research indicating environmental issues are of higher importance in industrialized countries, which have basic needs met, as compared to developing nations which may worry about food, clothing, or comfort (Dunlap et al. [3]). Importance of environmental issue varied across relevance due to need motivation and satisfaction. Another reason for variability across issue may be due to another basic need; health in particular. Credence has been given to the assertion that environmental problems are health threats to persons, thus tapping into the lower-level need of safety and security (Dunlap et al. [3]). Some environmental issues may be seen as higher-level and "luxury" concerns, while others, such as health, may be seen as basic concerns and viewed as relevant across all levels of socioeconomic status and national wealth. Further, persons believe poor environmental issues will negatively impact the lives of future generations (Dunlap et al. [3]), thus creating the need for improvement for safety and health issues. One other reason relevance may vary across environmental issue is media promulgation. Personally irrelevant environmental issues may not be salient in the mind of persons because of media exposure. Public concern regarding the environment increases when societal attention is paid toward the specific issues, even if the issue is not worsening (Dunlap et al. [4]). Further, there has been an emphasis on social facets of environmental issues, and less attention paid to the actual physical consequences themselves.

It also was hypothesized that a public safety variable would predict economic vitality. This hypothesis was supported, being that an increase in importance of public safety was associated with an increase in importance of economic vitality. There are several possible reasons public safety predicted economic vitality. The results insinuate that base needs, such as those for safety, remain of high importance, regardless of economic condition (UNLV Cannon Survey Center, Las Vegas Sun [17]). The results of the current investigation correspond with previous research. Specifically, according to Maslow [11, 12], persons may not concentrate on higher-level needs until lower-level needs are met. One issue for future research would be person-dependent levels of basic needs. For example, one person may require a higher-levels of safety compared to another in order to meet the requirements to ascend to higher-level needs such as those for elevated cognition (Sassoon [16]).

Some research suggests the entire way in which economies are measured and evaluated should be reconsidered. For example, Abramovitz [1] posits economies should compare a country's Gross Domestic Product (GDP) to its total material requirement as a measure of economic sustainability. Environmental stability, and in turn economic sustainability, would be a basic requirement necessary for persons to be capable of higher-level cognition and concern, such as those for ancillary environmental issues. One area of research should examine personally relevant economic issues. A relevant economic issue could be conceptualized as job growth, while national Gross Domestic Product (GDP) would not. A person dependent economic variable may be minimum wage. This variable may be relevant for those of low socioeconomic status but not for their higher counterparts. In accordance with this demographic variable, affluence and socioeconomic status may also have an effect on the relevance of variables. For example, government assistance programs may not be apposite for those higher in socioeconomic status, but of direct relevance to those lower in the construct. Ryan and Deci [15] suggest that once certain levels of wealth have been obtained, the feelings of security and stability have been met. At this point, more wealth adds little to increase well-being. Instead, the attainment of higher-level needs, such as psychological needs, should be more satisfying. Consistent with this reasoning Inglehart and Klingemann [7] found that happiness levels tend to be higher when average levels of income are high. Above a certain level of income however, happiness ceases to increase.

This study addressed one issue of methodological concern. Specifically, in past research participants have been asked about personal relevance directly, i.e. "How concerned are you personally about environmental problems?" (Dunlap et al. [3]). Participants were primed to think about the personally relevant issue, thus creating social desirability effects. Social desirability is a psychological mechanism used by persons in response to self-report items in which the respondent desires to be viewed in a positive light by others. Social desirability is an editing process by which persons think about potential responses to self-report questions, and then answer in a manner that will make them look favorable. This process often leads to less accurate responses (Holtgraves [6]). Past research has demonstrated social desirability in psychological studies using self-report surveys (Holtgraves [6]). Participants may have wanted to be viewed positively in the eyes of the researchers, thus giving more credence to the importance of the environmental issues. The current study examined the social construct of relevance, without the confounding variable of social desirability. To determine the relevant environmental issue, participants were asked to indicate the most urgent environmental issue affecting life, with the largest percentage stating water concerns. Participants were not primed to think about personal relevance, instead to think about urgency.

Like all research studies, this design is not infallible and a couple of problems with the study in particular should be addressed. The sample may have biases, thus creating external validity and generalization issues. Specifically, the sample may be considered biased because of the population in Nevada. The sample was not randomized and, therefore, could be considered only externally valid to Nevada residents. The study, however, should be considered valid because the bias was overcome by using a valid sample size with appropriate confidence levels with corresponding confidence intervals. One other issue is the specific variables. In particular, water conservation would not be applicable to other areas of the globe such as the Pacific Northwest in the United States due to varying climatic environments. The overall theory is considered to be valid, however, because the constructs of personal relevance were addressed. The specific variables would need further investigation in future studies of varying environmental conditions.

Several implications may be drawn from the current study. To begin, previous research would indicate persons are no longer concerned about environmental issues due to the far more important concern of economic stability. In actuality, persons may remain concerned about environmental issues, but only those with direct or vital relevance. Law makers, government, and the private sector can use this information in deciding importance and priority of issues. Relevance to persons can be used as an important construct in the decision making process.

Future research should investigate other hierarchical needs within the context of the current research design. For example, it is plausible that economic instability would create an importance of other lower order needs such as physiological, be it food, water, or comfort; as compared to social needs such as those for affiliation.

5 Conclusion

The extant literature indicates that in times of economic hardship, people afford less attention to environmental issues and, intuitively, more importance to economic matters. Further, persons are less likely to support environmental initiatives during tough economic times on an individual level, and some research shows that there is a difference between importance of national economy and environmental issues, with the latter being less imperative (Franzen and Meyer [5]; Lee and Norris [8]). Environmental issues that are more personally relevant remain important in times of economic hardship, while others that are less individually relevant would be less important. In addition, persons may be more concerned about basic needs, such as those for public safety, compared to environmental issues. This would be particularly true for environmental issue containing less personal relevance. The current paper provided an empirical research investigation analyzing the relationship between perceptions of economic vitality in relation to environmental relevance and public safety.

References

[1] Abramovitz, J. N. Report turns ‘economic-versus-environment’ argument on its head. *World Watch*, *10*(5), 9, 1997.

[2] Anderson, K. Economic Growth, Environmental Issues, and Trade. C.E.P.R. Discussion Paper, CEPR Discussion Papers: 830, 1993.

[3] Dunlap, R. E., Gallup, G. H., & Gallup, A. M. Of Global Concern: Results of the Health of the Planet Survey. *Handbook of Global Environmental Policy and Administration.* Eds. D. L. Soden and B. S. Steel. Basel, Switzerland: Marcel Dekker, 1999.

[4] Dunlap, R. E., Lutzenhiser, L. A., & Rosa, E. A. Understanding Environmental Problems: A Sociological Perspective. In Economy, Environment, and Technology: A socio-economic approach. Ed. B. Burgenmeier. Studies in Socio-Economics, N.Y: Sharpe, 1994.

[5] Franzen, A., & Meyer, R. Environmental attitudes in cross-national perspective: A multilevel analysis of the ISSP 1993 and 2000. European Sociological Review, doi: 10.1093/esr/jcp018, 2009.

[6] Holtgraves, T. Social desirability and self-reports: Testing models of socially desirable responding. *Personality and Social Psychology Bulletin, 30*, 161-172, 2004.
[7] Inglehart, R., & Klingemann, H. Genes, culture, democracy, and happiness (pp. 165-183). In E. Diener and E. Suh, (Eds.) Culture and Subjective well-being. Cambridge, MA: The MIT Press, 2000.
[8] Lee, A, R., & Norris, J. A. Attitudes toward environmental issues in east Europe. International Journal of Public Opinion Research, 12(4), 372-397, 2000.
[9] Kakrida, U. V. Strong links between the economy and the environment, *Review of Economic Sciences, 12,* 67-84, 2007.
[10] Maslow, A. H. *Motivation and personality*. New York: Harper & Row Publishers, 1987.
[11] Maslow, A. H. *The farther reaches of human nature*. Oxford, England: Viking, 1971.
[12] Maslow, A. H. *Toward a psychology of being*. 2nd ed. Oxford, England: D. Van Nostrand, 1968.
[13] Okina, K. Concluding Panel Discussion: Sustained Economic Growth and Central Bankning: Introductory Remarks. *Monetary and Economic Studies (Special Edition) 24,* 217-224, 2004.
[14] Perelman, M. The perverse economy: *The impact of markets on people and the environment*. New York and Houndmills, UK: Palagrave Macmillan, 2003.
[15] Ryan, R. M., & Deci, E. L. On happiness and human potentials: A review of research on hedonic and eudaimonic well-being. *Annual Review of Psychology, 52,* 141-166, 2001.
[16] Sassoon, J. Self-actualization: Theory and technology. Pointe Claire, PQ, Canada: Humanica Press, 262, 1988.
[17] UNLV Cannon Survey Center, Las Vegas Sun, 2009.

Values held by young stakeholders on financial planning regarding liabilities for nuclear decommissioning

B. Labor[1] & S. Lindskog[2]
[1]*Badania Dydaktyczne, Poland*
[2]*The Swedish Radiation Safety Authority, Sweden*

Abstract

Nuclear waste programmes have from time to time encountered set-backs related to lack of trust from society. One experience is that it is necessary that all nuclear work be conducted in accordance with high standards on health and the environment, including funding for future environmental liabilities according to the Polluter Pays Principle (PPP). Another experience is that it is imperative to communicate with the stakeholders, and to understand the reasons for their positions.

Surveys on opinions have usually been based on response by letter and similar. In particular for young stakeholders, this has implied very low response rates and associated low validities of the results. The present work is based on an experimental programme with altogether 880 personal interviews in three towns in Poland with a near 100% response rate.

The conclusions include the following:

- sustainable energy sources are favoured (nuclear power is one option);
- the values are based on safety and environmental aspects;
- the Polluter Pays Principle is to be honoured;
- there is a scepticism concerning the will and capability of present decision makers to implement this principle;
- there is too little basis for educated opinions;
- there is no difference between the sexes;
- it is necessary to use more comprehensive communication channels.

Keywords: nuclear, decommissioning, dismantling, radioactivity, environmental, liability, fund, cost calculation, stakeholders, values.

WIT Transactions on Ecology and the Environment, Vol 131, © 2010 WIT Press
www.witpress.com, ISSN 1743-3541 (on-line)
doi:10.2495/EEIA100201

1 Background

A large number of nuclear facilities - including nuclear power plants, research reactors and various other nuclear facilities - are approaching the end of their economic as well as technical life times, and will therefore enter the decommissioning phase during the present decade. Such facilities are contaminated with radioactive material and frequently also activated, and their decommissioning represents huge environmental liabilities. It is therefore pertinent that sufficient financial resources to cover all future costs be accumulated by the generation reaping the benefits from nuclear power production.

The task to make allowances for future costs is governed by the Polluter Pays Principle (PPP). However, the calculations must be of such quality that they are generation-neutral, and funding compliant to the Extended Polluter Responsibility (EPR). It is occasionally put forward that conventional economic analysis tends to treat future generations unfairly [1], i.e. tends to underestimate future costs. Complex environmental liability situations include long operation and decommissioning times and often concern more than one generation. Here, implementation of the EPR principle implies that the "property rights" of future generations must be fully respected and acknowledged in the financial systems for funding of decommissioning and dismantling of nuclear facilities. One key step to meet this daunting task is to safeguard and assure that appropriate financial resources are accumulated in segregated funds such that no financial liabilities are passed on to future generations.

In Sweden, for example, such an approach has been formalised in a new ordinance that came into force on November 1, 2008. It states that all environmental liabilities for future decommissioning and associated waste management must be assessed and declared, and that funding must be secured to cover the estimated future environmental liabilities. The estimation must be in concordance with good accounting practice. This means that, in general, the error in the estimates may diverge by not more than ± 15% [2].

2 The strive for enhanced efficiency

The long term process to manage the decommissioning of nuclear facilities and the associated waste can be seen as a general efficiency maximisation problem that is subject to constraints. These constraints comprise not only conventional techniques, including associated contemporary research and development, but also include social questions and especially financial and democratic issues. The social aspects need to be addressed in an open and transparent manner in order to enable a sound earning of support from the general public. Such a support should be based on a sustainability and long term credibility of nuclear waste liabilities as well as planning and implementation of decommissioning and dismantling of nuclear facilities.

The key to success in defining and establishing such a system, in which these constrains are fully considered, is earning long-term support from the general public. The strategic importance of this issue was realised when commercial nuclear power was introduced in Sweden. This lead to the issuing of the Financing Act, which with only moderate adjustments is in force still today.

One major threat to the system is that costs may exceed what has been budgeted and funded. It has proven notoriously difficult to obtain reliable estimates, primarily due to specific cost driving factors linked to critical features and events.

In some other countries, major set-backs have occurred so that the nuclear waste programmes have been delayed, redirected or halted. Such events are of course related to the presence of well functioning democratic processes, and are also of critical importance for achieving full legitimacy for the implementations of proposed methods and techniques for nuclear waste management.

From the perspective of financial control and efficiency, such set-backs may lead to consumption of more financial resources than in an optimal case. Adverse developments might be mitigated by improving the quality of the cost estimates and the planning process such that set-backs are either controlled or reduced to a minimum. From the perspective of efficiency, there is also strong support to accept and implement the Extended Polluter Responsibility (EPR) principle so that the inter alia rights of future generations are taken into consideration already from the early phases of cost estimating and decommissioning planning. In this process, the inclusion of the younger generation can be seen as a resource that can contribute to a good efficiency.

3 The inclusion of the public

The Swedish system may be used as an example of how enhanced public inclusion and how good application of associated methods (e g for conflict reduction) may contribute to more efficient processes. During the last 20 years there have been some major activities for public dialogue and risk communication between stakeholders [3-6]. The Swedish system has, for instance, given non-governmental organisations opportunities to participate in the planning for final disposal, and thereby broadened the democratic base of the project. One common weakness of these projects is, however, that they have not been based on the Extended Polluter Responsibility (EPR) principle, and therefore inter alia have not been successful in including younger generations.

It is generally accepted that decommissioning strategies will vary according to a number of considerations. Hence, it is for example anticipated that the accessibility and availability of waste disposal alternatives, the quality of radiological mapping, the radiation protection policy, cost and funding considerations and local site factors are highly significant for the process of defining the general prerequisites for efficient decommissioning strategies. In this context, the significance of the choice of methodology for communication of decommissioning strategy needs to be identified and evaluated at a first step, see for example [3, 4, 7 (pp. 9-19)]. The aim is to reduce the risks for delays in the decommissioning phase and thus to alleviate challenges which would otherwise demand increased expenditure and specialist knowledge as the conditions of facilities decay and deteriorate, see for example [7 (pp. 7-19)-9].

The collection of comprehensive and authentic data about the values of younger stakeholders may help the present generation to bridge the gap between the planning of the present generation for the long term safe management of nuclear waste and the financial abilities of future generations to implement the plans. Such data and knowledge may, in turn, help us to find better and more appropriate socially constructive forms for inclusion of younger stakeholders at an early stage. This will enhance our abilities to design platforms for sustainable communication with all stakeholder groups. It may also contribute to enhancing the efficiency in the planning process and final decommissioning decisions since the younger stakeholders of today will most likely be the actual decision makers. The failure to incorporate the views of younger stakeholders in the planning and cost calculations may well prove to be a cost raiser. It is prudent practice to base current cost calculations on such robust assumptions that ensure that any myopia of the present generation is taken into consideration. For such prudent practice, it is pertinent to compile and make use of the values of younger stakeholder.

4 The values of younger citizens

The intrinsic properties of projects of decommissioning and dismantling with their trans-generational characteristics and associated longitudinal base call for an early inclusion of stakeholders. The interaction with groups in the local municipality is highly significant for the successful planning and completion of a project. Issues such as overall political prerequisites and national interests call for an open and transparent flow of information also on the national level.

The inclusion of all stake holder groups is essential. It must be stressed, however, that the opportunity to include the younger citizens at an early stage in the democratic decision making process provides a unique possibility to gain commitment and support from a future group of stakeholders already today. In this way, unnecessary and time consuming reproduction of processes can be avoided.

In the past, our ability to make conclusions in this area has been severely limited by the low levels of response frequencies. In addition, the answers have been inconsistent, thus reflecting disturbances in the communication process.

One common feature of the previous studies is the use of the same batteries of questions together with identical layouts of the questionnaires for all age groups. This may well explain the low response frequencies as well as the questionable quality of the answers. It is well known that young persons communicate by SMS and e-mail rather than letters in the ordinary mail, as opposed to older people. It can thus be readily predicted that the response frequency will be low for young people. On the other hand, if a more experimental model for data gathering is used for the younger stakeholders, it can be predicted that the response rate is higher [10]. Hence, such considerably higher response frequencies can be expected in cases where data retrieval is done in authentic environments where advantage is taken of comprehensive knowledge about learning processes.

Authentic data for Swedish conditions from a county with nuclear power have been collected in an explorative study in Kalmar in 2007 [11]. It can be expected, however, that the attitudes in a county with nuclear power may differ substantially from those of a county without nuclear power.

This need for more generally valid information has prompted the present survey which is from the neighbouring country Poland. It has no nuclear power plants, but a vivid national dialogue on building such plants. The conditions in Poland are thus close to ideal for obtaining "bench-marked" data of the values of younger citizens regarding decommissioning for the purpose of comparison.

5 The survey data

The scope of the present paper is to present some results from the above mentioned survey based on an experimental design, and to analyse this material with regard to the attitudes of younger people towards decommissioning and nuclear waste [12, 13]. The data was collected as a stratified sample during January to November 2008 in the Polish towns of Gdansk, Lublin and Elblag, and the numbers of responses are shown in Table 1.

Table 1: The number of responses in the survey.

City Gender	Gdansk	Elblag	Lublin	Total	%
Men	249	38	105	392	50,30%
Woman	119	89	180	388	49,70%
Total	368	127	285	780	100,00%

6 The Swedish case

The Ågesta electricity and district heat producing heavy water nuclear power reactor in the outskirts of Stockholm was shut down in 1973. The two light water nuclear power reactors at Barsebäck near Malmö were shut down in 1999 and 2004, respectively. The large (50 MW_{th}) research reactor at Studsvik, 100 km south of Stockholm, was shut down in the year 2005. At present, the decommissioning of these reactors is in a transition phase pending access to space for final disposal of long-lived low and intermediate level waste which is anticipated to be available at the beginning of the 2020's.

Thus, the extent of the transition period, i.e. the time elapsed between the final shutdown and the last day of the dismantling, can be estimated to be at least one generation.

This implies that it will be the next (or some successive) generation that will be responsible for the decommissioning and the associated waste management. Thus, the younger citizens are, and will be, stakeholders for a long time. In an increasingly unified Europe, this may concern not only Sweden but also (what is presently) other countries.

7 Summary of the results from the survey

A summary of the findings of the survey is presented below in Sections 7.1 – 7.7. Details of the study can be found in [12, 13]. A general finding is that the differences between the sexes as well as those between the three towns are insignificant.

7.1 The energy mix preferences

In the aggregated survey data, a total of 780 students gave 1123 answers on the multiple answers question: *Which form of energy do you prefer?* In total, 35% declared that they prefer windmills. Hydro power was said to be preferred by 32% followed by nuclear power that was regarded as preferable by 25%. Energy produced by coal was seen as an acceptable option for as little as 2,8%.

The survey data supports the view that younger stakeholders are in favour of renewable and long-term sustainable energy sources vis-à-vis non-renewable counterparts. Here it is perceived that hydro power and wind power, e.g. windmills, are the two most popular energy sources. At the same time it may be envisaged that as much as ¼ are accepting nuclear power as a future potential energy supply mode. Further details are presented in Table 2.

Table 2: Which form of energy do you prefer?

	City	Gdansk		Elbląg		Lublin		Sub SUM		SUM	
Answer	Gender	M	W	M	W	M	W	M	W	MW	%
Coal		8	13	1	4		5	9	22	31	2,8
Nuclear Power		72	19	18	16	78	80	168	115	283	25,2
Hydro Power		106	44	16	40	60	97	182	181	363	32,3
Windmills		111	61	21	53	52	99	184	213	397	35,4
Miscellaneous		9	4		2	16	18	25	24	49	4,4
Total		297	137	56	113	206	299	568	555	1123	100

7.2 The implicit application of the Polluter Pays Principle

A total of 786 students answered the question: *Who shall take care of the Swedish nuclear waste?* As many as 88% answered that they thought that Sweden should take care of its nuclear waste. This answer clearly indicates that the younger stakeholders have a firm belief in the Polluter Pays Principle (PPP).

Only 12% of the students held the opinion that the nuclear waste generated in Sweden might be managed also elsewhere within the European Union. Such a view does not necessarily imply a breach of the PPP principle since the polluter might still pay the full costs.

The conclusion is that the data strongly indicates that the Polluter Pays Principle has a solid foundation among the younger European citizens.

7.3 Younger stakeholders and the decision makers

On the question: *Do you have confidence and trust in the decisions makers' capability in the decommissioning process?* the distribution of the answers was an almost perfect image of an Erlangen distribution, see [12].

In the cluster samples 18% gave the response that they trust the decision maker's determinations and commitments as well as their ability to solve the problem of final storage of radioactive waste from decommission of nuclear power plants, including planning, construction and operation. Furthermore, 26% gave the answer that they had some kind of mistrust in the decision maker's commitments on this matter. The bulk of respondents, 57%, declared that they did not have a clear position in this matter, and that they were unsure about how to respond. Further detail is provided in Table 3.

Table 3: Do you have confidence and trust in the decisions makers' capability in the decommissioning process?

	City	Gdansk		Elbląg		Lublin		Sub SUM		SUM	
Answer	Gender	M	W	M	W	M	W	M	W	MW	%
Yes		37	29	8	8	34	23	79	60	139	17,8
Do not / unsure		135	58	19	56	55	120	209	234	443	56,7
No		77	33	11	25	16	37	104	95	199	25,5
Total		249	120	38	89	105	180	392	389	781	100

7.4 Propensity to live in the vicinity of a final storage for nuclear waste

The reason to pose the question, *Can you consider having a site for final disposal of nuclear waste near to your home?* to the respondents was to evaluate the propensity to live in the vicinity of a nuclear waste final storage facility. Less than 20% - or 152 out of a total of 781 of the students - said that they could consider this. Consequently, over 80% were against having a nuclear waste site near to their homes.

7.5 The platform of the younger stakeholder's opinions

The values that the younger stakeholders' base their views on are summarised in Table 4. In the samples, a total of 15% expressed that their opinions were based on trust for the stakeholders involved. Another 12% said that their opinions were based on the opportunities, both financial and social, linked to major decommissioning projects. Almost one third, 29%, expressed that their opinions were based on insufficient knowledge. More than one third, 38%, said that their opinion was based on uneasiness about the risks involved. A small group comprising 6% gave other reasons for their views.

The data shows that around two thirds of the younger citizens base their values on the risks associated with the handling of nuclear waste together with lack of knowledge.

The value base also includes trust for the involved stakeholders which accounts for around one fifth to one sixth of the explanation as well as opportunities linked to a disposal of nuclear waste for around one tenth of the explanation value.

The data suggests that younger citizens base their values mainly on fundamental questions concerning risks and lack of knowledge (i.e. nuclear waste information), and less on more opportunistic issues like future benefits to the region from a nuclear waste storage and trust for the stakeholders involved in the decision making process.

7.6 Crucial aspects for the younger stakeholders' acceptance of a final repository for nuclear waste

In the three regional samples, 38% said that the safety aspect is the most important one for considering a repository for nuclear waste. The environmental

Table 4: What do you base your opinions on?

	City	Gdansk		Elbląg		Lublin		Sub SUM		SUM	
Answer	Gender	M	W	M	W	M	W	M	W	MW	%
Trust for the stakeholders		40	16	3	9	35	34	78	59	137	14,9
Nuclear waste disposal opportunities		24	7	9	8	31	29	64	44	108	11,7
Lack of knowledge		86	54	13	28	21	67	120	149	269	29,2
Unease with risks		111	49	14	51	32	93	157	193	350	38,0
Miscellaneous		19	6	4	1	11	15	34	22	56	6,1
Total		280	132	43	97	130	238	453	467	920	100

Table 5: What aspects are most important for acceptance?

	City	Gdansk		Elbląg		Lublin		Sub SUM		SUM	
Answer	Gender	M	W	M	W	M	W	M	W	MW	%
Safety aspect		87	5	26	61	97	166	210	232	442	38,0
Environmental aspect		63	2	17	45	63	119	143	166	309	26,5
As far as possible from home		59	5	14	44	45	97	118	146	264	22,7
Methods and techniques		19		9	4	25	52	53	56	109	9,4
Economic growth		7		6	3	6	7	19	10	29	2,5
Miscellaneous		1		1		5	4	7	4	11	0,9
Total		236	12	73	157	241	445	550	614	1164	100

aspect was seen as the second most important factor with a total answer frequency of 27%. The third most important factor was whether the localization of the disposal facility should be as far from home as possible which was put forward by 23%. These three reasons have an explanatory power of nearly seven eights, or more exactly 87%, see Table 5 for further detail.

7.7 The information channels that are prioritised by the younger stakeholders

Data on the preferences of younger stakeholders on communication channels and information processes with regard to nuclear waste is presented in Table 6.

The respondents also gave a lot of suggestions for possible improvements and developments regarding the shape and form of information in the areas of decommissioning of nuclear facilities and management of nuclear waste:

- It may be appropriate to notice that the alternative to participate in project groups (collective learning) was the most popular alternative that was suggested in 29% of the cases.
- The alternative of producing a film was mentioned in 19% of the cases.
- The suggestion to plan, organise and conduct exhibitions was given in 16% of the cases.
- The alternative to use, create and construct websites was given in 15% of the case.
- The suggestion to use "power point" presentations was suggested in 13% of the cases.
- The alternative to make a theatre play was an option for 4,9% of the respondents.

8 Some tentative conclusions

The data presented above supports some tentative conclusions on the values of younger European citizens regarding decommissioning of older nuclear facilities and disposal of the nuclear waste. The main findings are as follows:

Table 6: Channels for communication preferred by younger stakeholders.

	City	Gdansk		Elbląg		Lublin		Sub SUM		SUM	
Answer	Gender	M	W	M	W	M	W	M	W	MW	%
Project work		55	3	19	43	71	141	145	187	232	29,3
Exhibitions		26	5	7	23	41	83	74	111	185	16,3
Power-Point presentation		28	3	7	18	32	56	67	77	144	12,7
Create web pages		29	3	13	29	39	54	81	86	167	14,7
Theatre		6	1	4	13	8	24	18	38	56	4,9
Film		58	4	17	36	28	71	103	111	214	18,9
Other		4		2	1	13	15	19	16	35	3,1
Total		206	19	69	163	232	444	507	626	1133	100

- Younger citizens tend to favour long term sustainable energy sources, e.g. hydropower and windmills, before non-sustainable techniques, e.g. nuclear power and carbon based fossil fuels.
- Younger citizens tend to base their values on safety and environmental aspects, while other aspects like contribution to future economic growth in the society and development of technological processes are less influential on the values.
- Younger citizens have a limited trust, or scepticism, concerning the capabilities and skills of the contemporary decision makers to develop and present prudent, sound and long term sustainable management of nuclear waste from decommissioning of nuclear facilities.
- Younger citizens tend to refer to a deficit of knowledge, information and debate as a base for their judgement of the decision makers' skills, determination and credibility on this topic.
- Younger citizens can demonstrate major and well defined suggestions on how more information can be made accessible to the general public. They are also able to present a wide spectrum of methods, ways and modes to include younger people in the topic and the related democratic processes.
- Younger citizens have the propensity to prioritize aspects such as safety and environmental questions above financial aspects and economic growth as criteria for distribution and allocation of different sources of energy.

9 Findings and recommendation: a synthesis

Our study supports the statement that the younger generation is likely to support not only the general Polluter Pays Principle and the major thoughts in the European environmental codex, but are also interested in participating and giving contributions to open democratic processes regarding ways in which the nuclear waste of today shall be managed. They are open to communication and are accepting a wide selection of communication channels. Hence, the younger stakeholders can fulfil their part of the Extended Polluter Responsibility (EPR).

Interestingly, and in contrast to results obtained previously for older persons, we have found no significant differences between the sexes in the values and attitudes.

An early entrance in the planning phase is normally seen as one prerequisite for a successful dismantling and decommissioning of permanently closed nuclear facilities. Likewise, an early entrance of the younger citizens in the planning for dismantling of older nuclear facilities and management of nuclear waste is likely to have a positive effect on the accuracy of the cost calculations of the future financial liabilities, and thereby also on the robustness of the entire process.

Hence, it appears reasonable to encourage that younger citizens already at an early stage engage in an open and democratic decision process concerning decommissioning of nuclear facilities and the associated waste management. This will improve the planning process and at the same time the quality of the cost calculations. This will, in turn, contribute to better use of eared-marked resources for decommissioning and thereby reduce the risk that future generations might face a deficiency in terms of a deviance between actual costs and funds [14].

Conversely, failure to involve the younger citizens in the process may result in cost driving activities, since a longer period of time will be needed in order to go through the decommissioning process. Furthermore, a prolonged process will also increase the total detriment from health, safety and environmental perspectives. Thus, failure to include younger stakeholder is a potential cost riser.

The presented data demonstrates that younger stakeholders can be successfully included in the process if appropriate channels for communication are selected based on the preferences of this stakeholder group. Thus, risk communication is not only a question of the content but also a question of choosing appropriate channels for communication. The possibilities for fruitful communication can be enhanced if the nuclear waste information is based on comprehensive didactic methods specially designed for younger age groups.

References

[1] Padilla, E., *Intergenerational equity and sustainability.* Ecological Economics 41 (2002) 69-83.

[2] Lindskog, S. & Sjöblom, R., *Radiological, technical and financial planning for decommissioning of small nuclear facilities in Sweden.* Proceedings of the 12th International Conference on Environmental Remediation and Radioactive Waste Management, ICEM2009, October 11-15, 2009, Liverpool, UK.

[3] Espejo, R., *Structure for Transparency in Nuclear Waste Management, Comparative Review of the Structures for Nuclear Waste Management in France, Sweden and the UK.* SKI* Report 2003:26, November 2002, p. 30.

[4] Andersson, K., et al, *Design and Evaluation of Public Hearings for Swedish Site Selection, A Report from the RISCOM II Project.* SKI* Report 2003:3. August, 2003.

[5] Andersson, K., et al., *Public Values and Stakeholder Involvement – A New framework for Performance Assessment, A Report from the RISCOM II Project.* SKI* Report 2004:24. October, 2003.

[6] Elam, M. and Sundqvist G., *Stakeholder Involvement in Swedish Nuclear Waste Management.* SKI* Report 2007:02. September, 2006.

[7] Laraia, M. & McIntyre, P. J., responsible officers; Cato A., Lindskog S. & Sjöblom. R. et al contributors. *Decommissioning of research reactors and other small facilities by making optimal use of available resources.* IAEA Report Series 463, Vienna 2008.

[8] *Geological Repositories: Political and Technical Progress.* OECD, Nuclear Energy Agency (NEA). OECD 2005, pages 61-65.

[9] Frantisek, O., et al., *Methodology for description and evaluation of cost profiles and cost structure during transition period.* SKI* Report 2008:38. May, 2008

[10] Sjöberg, L., *Attityd till slutförvar av använt kärnbränsle.* (Attitudes to a final repository for spent nuclear fuel. In Swedish. Swedish Nuclear Fuel and Waste Management Company, SKB. SKB Report R-08-119. September 2008.
[11] *Skolungdomars inställning till slutförvar av kärnavfall.* (The attitude of students to final storage of nuclear waste. In Swedish). The Regional Council of Kalmar. Kalmar, Sweden, 2007.
[12] Tyszkiewicz, B. and Labor, B., *A Survey of Younger Citizens Values towards Decommissioning and Dismantling of Older Nuclear Facilities in a European Perspective.* SKI* Report 2008:52. June, 2008.
[13] Tyszkiewicz, B. and Labor, B., *On Younger Stakeholders and Decommissioning of Nuclear Facilities.* SSM* Report 2009:32. August, 2009.
[14] Lindskog, S., Cato, A. & Sjöblom, R., *Estimations of costs for dismantling, decommissioning and associated waste management of nuclear facilities, and associated impact on decision processes, functioning of markets and the distribution of responsibilities between generations.* Environmental Economics II, 28-30 May 2008, Cadiz, Spain. WIT Transactions on Ecology
* SKI = Swedish Nuclear Power Inspectorate, now Swedish Radiation Safety Authority (SSM). Reports are available at www.ssm.se.

Evaluating the complementarity of the educational function in agriculture

Y. Ohe
Department of Food and Resource Economics, Chiba University, Japan

Abstract

Agriculture generates positive externalities to society in addition to food production without receiving compensation, which is called the multifunctionality of agriculture. The educational function of agriculture has not been fully examined despite the growing need for such services from urbanites. This paper focused on the educational function of agriculture by considering it as a joint product of farm production in terms of not only technical complementarity, but also of institutional complementarity, which includes institutional aspects such as food and rural culture. Data were obtained from the URL of the Farm Experience Net, launched in May 2003 and operated by the National Association for Promotion of Rural Youth Education in Japan. We employed statistical tests to clarify the features of the bodies or individual farmers who provide experience services. Findings and implications were as follows. Firstly, from the perspective of complementarity, we can classify that crop-wise or operation-wise experiences showed a strong connection with the technical complementarity of farms with seasonality, while food and rural cultural experiences had a strong connection with institutional complementarity with less seasonality. Secondly, non-governmental organizations, the quasi-public sector and farmers' groups played important roles in food and rural culture experience services. Thus, we should recognize the significance of both technical and institutional complementarity to extend the educational function of agriculture and the rural heritage. The combination of the two types of complementarity is effective in building a wide range of networks of those stakeholders who provide food and rural experience services while easing the constraints of seasonality.
Keywords: technical complementarity, institutional complementarity, educational function of agriculture, multifunctionality of agriculture, food and rural experience services, farm diversification, cultural diversity.

WIT Transactions on Ecology and the Environment, Vol 131, © 2010 WIT Press
www.witpress.com, ISSN 1743-3541 (on-line)
doi:10.2495/EEIA100211

1 Introduction

The educational function of agriculture is considered as a part of multifunctionality (Science Council of Japan [6]). Multifunctionality is a joint product with externality (OECD, [2, 3]). Nevertheless, the process that generates multifunctionality has not been limited to cases of technical complementarity. Hagedorn [1] pointed out the existence of institutional complementarity as the institutional jointness in addition to technical complementarity. In the case of institutional complementarity, joint products are generated through the institutional process and factors. It is not uncommon in every country that the process of agricultural production is determined by institutional factors that have been formed historically for centuries. So, it is necessary to consider both technical and institutional complementarity to explore the issues of complementarity in multifunctionality in agriculture. Particularly, institutional complementarity plays an important role in generating multifunctionality in terms of social and cultural function. The educational function is such an example. Ohe [4] evaluated multifunctionality by focusing on the village function by incorporating the concept of institutional complementarity. Ohe [5] also evaluated the significance of educational dairy farms in connection with institutional complementarity. However, to my knowledge, as yet a full-fledged conceptual and empirical examination of the aspect of the connection between institutional complementarity and the educational function has not been conducted.

While farming experiences are connected with technical complementarity, experiences involving the food and agrarian heritage that include an educational function have a connection with institutional complementarity. To understand the educational function of agriculture properly, it is safe to say that we cannot overlook the role of institutional complementarity. Therefore, here we evaluate institutional complementarity from the agrarian cultural point of view. First, we make conceptual considerations of technical and institutional complementarity and then empirically examine their characteristics. Finally, we give policy recommendations on how to promote the educational function of agriculture in the future.

2 Two concepts of complementarity

The most common image of complementarity is the case of technical complementarity that represents the technically determined process of production of by-products that accompany the main products. The joint production of mutton and wool is a typical textbook example of technical complementarity. Traditionally, farm households, however, have been involved in not only agricultural production but also in producing processed food and crafts that richly reflect the distinctiveness of the local area. These products have been important elements that constitute local food culture and agrarian heritage. In this respect, these by-products are products of the institutional complementarity of food and agrarian heritage. Table 1 shows a comparison between technical and institutional complementarity.

Table 1: Comparison of technical and institutional complementarity.

Type	Process of generating complementarity	Experience services
Technical complementarity	Technical process (agricultural production process)	Crop-wise services Operation-wise services
Institutional complementarity	Institutional process (rural life process)	Food culture services Rural culture services

Technical complementarity is basically determined by technical conditions and it is not manageable for farmers in the short/medium term. This fact, however, does not mean that technical complementarity is not variable. Agricultural production in reality, particularly crop farming, has apparent seasonality. For instance, rice planting and harvesting are the most common farming experiences and are naturally constrained by seasonality. In short, technical complementarity regarding the educational function involves seasonality in agricultural production.

Likewise, institutional factors, such as customs and culture that have been traditionally formed and constitute institutional complementarity, are not manageable in the short/medium term. These institutions, having traditionally been the foundation of agrarian life and production activities, are not selective, but are commonly applied to every member of the local community and are difficult to change. Thus, institutional complementarity enables us to have a wider scope in understanding the significance of local cultural diversity and farm diversification, which is not always the case with technical complementarity.

3 Data and methodology

Data were obtained from the URL of the Farm Experience Net, launched in May 2003 and operated by the National Association for Promotion of Rural Youth Education in Japan, a body under the administration of the Ministry of Agriculture, Forestry and Fisheries (MAFF). Data were collected from March to April in 2003 from the URL.

The merit of these data is that they have the widest coverage of farm experience services throughout the country, providing information on by whom, when, where and how experience services are provided. However, a demerit is that the data are devoid of factor input information necessary for the examination of factor input relationships such as production or the cost function. We employed statistical tests to clarify the features of the bodies or individual farmers who provide experience services. The sample size was 385 providers.

Experience services are classified into four types: crop-wise experiences, farming operation experiences, food culture and rural culture (Table 2). We identified the five most popular items for each type of experience service. Among crop-wise experiences, vegetables come first, followed by rice, dairy, fruit, and wheat/buckwheat. Operation-wise, the order of popularity was harvest,

Table 2: Breakdown of experience services.

Crop-wise services: vegetables (191), rice (151), dairy (129), fruit (121), wheat/buckwheat (114), livestock (88), flowers (60)
Operation-wise services: harvest (250), planting (186), cultivation (174), taking care of livestock (131), milking (107), soil improvement (100)
Food culture services: making soba and udon noodles (109), rice cakes (107), dairy products (71), pickles (69), natto/tofu (56), fruit juice (38), mannan (36), miso (16), jam (16), sausage (11), bread/cookies (14), local cuisine (12), herb tea (3), others (104)
Rural culture services: harvest festival (105), straw craft (69), fire fly catching (48), morning bazaar (34), charcoal-making (18), collecting edible wild plants/picking mushrooms (17), bamboo craft (16), woodcraft (15), dyeing (15), pottery (11), vine craft (8), wreath-making (7), watching fish (2), others (135)

Source: the URL of the Farm Experience Net, owned by the National Association for Promotion of Rural Youth Education in March 2005. Total sample size is 385.
Note: Figure in the parenthesis means sample size.

Table 3: Partial correlation coefficients among experience services.

Services	Crop-wise	Operation-wise	Food culture	Rural culture
Crop-wise	1			
Operation-wise	0.4682***	1		
Food culture	0.3824***	0.3433***	1	
Rural culture	0.2351***	0.2308***	0.6268***	1

Source: Same as Table 2.
Note: *** 1% significance.

planting, cultivation, taking care of livestock, and milking. With regard to food culture experiences, experiences in preparing a variety of foods are provided, with buckwheat and udon noodles, rice cake, dairy products, pickles and soybean products (i.e. natto, fermented beans, and tofu-making) accounting for the five most popular. Of the wide range of activities under the topic of rural culture experiences, the five most often offered are harvest festivals, straw crafts, firefly catching, morning bazaars and charcoal making. Generally, the closer to the production process of agriculture, the more tightly the technical complementarity becomes. In this context, we can define that crop-wise and operation-wise experiences represent technical complementarity while food and rural culture experiences represent institutional complementarity because the agrarian heritage particularly works on these two culture experiences.

Regarding the statistical relationships shown in Table 3, the four types of services have statistically significant positive correlations with each other; this is especially evident between crop-wise and operation-wise services (partial coefficient 0.468 with 1% significance) and between food and rural culture services (0.627 with 1% significance). In the case of institutional complementarity, non-farm bodies can become providers, so we will empirically examine characteristics of each complementarity taking into account the differences in providers.

4 Empirical examination on the complementarity of farm and rural experience services

4.1 Seasonality of experience services

First, we examined the seasonality of the four types of services by focusing on the portion of providers in year-round operations. The result shows that there was a difference with statistical significance in crop-wise and operation-wise experiences as tabulated in Table 4 while there was no significant difference between food and rural culture experiences except for dairy products with 10% significance.

Among crop-wise services, the portion of providers with year-round operations was more than 60% of those providing flower and dairy farming experience services but was less than 50% of those offering fruit and vegetable experiences, reflecting the stronger production seasonality of the former (each 5% significance). In operation-wise services, that portion was more than 60% for dairy farming and livestock husbandry and less than 50% for crop farming, indicating stronger seasonality in the latter (each 1% significance).

To summarize, we statistically confirmed that dairy farming has weaker seasonality than crop farming whereas cultural experiences have negligible

Table 4: Percentage of year-round operations in crop-wise and operation-wise services.

Crop-wise experience services							
Crop	Fruit	Vege-tables	Rice	Wheat, buckwheat	Livestock except dairy	Dairy farming	Flowers
Sample size	121	191	151	114	88	129	60
% year-round operation	42.2**	48.7	55.0	57.0	59.1+	63.6***	65.0**
Operation-wise experience services							
Operation	Harvest	Culti-vation	Planting	Soil improvement	Rearing management	Milking	-
Sample size	250	174	186	100	131	107	-
% year-round operation	45.6***	47.1+	48.9	58.0+	61.1***	64.5***	-

Source: same as Table 2.
Note: Chi-square test was conducted; ***, **, *, + indicate 1%, 5%, 10%, 20% (reference) significance, respectively.

seasonality. Therefore, we can say that the combination of the two types, farm operation experiences with technical complementarity of tighter seasonality and culture experiences with weaker seasonality, will enable providers to offer year-round services.

4.2 Provision of experience services

Providers are classified into three types according to principles related to profit-making behaviour: family farms and corporate farms aiming at profit maximization belong to the private sector (46.7%); national and prefectural governments, municipalities and research and educational institutions are considered as the public sector (24.7%); and intermediate bodies such as NGO, voluntary associations and agricultural cooperatives comprise the quasi-public sector (28.6%). Thus, non-private providers account for more than half of the total providers of farm and rural experience services (53%) (Figure 1). One reason is that the public sector promotes experience services proactively as a way of policy implementation, so we should consider this policy bias with regard to the relatively higher portion of public and quasi-public providers. These public and quasi-public providers offer their own experience services and play a role in organizing a network of providers from the private sector, such as farmers. Put differently, the public sector has two roles: as actual providers of services and as organizers of providers' networks. In this sense, it should be said that there is role sharing in educational services among sectors. Even services provided by the public sector are under the influence of the two types of complementarity.

Of crop-wise services, as shown in Table 5, fewer were offered in family farms (2.1 services), with more services being offered by the public sector (3.2 services). In contrast, regarding food and rural culture services, the quasi-public sector provided more services while the public sector provided fewer.

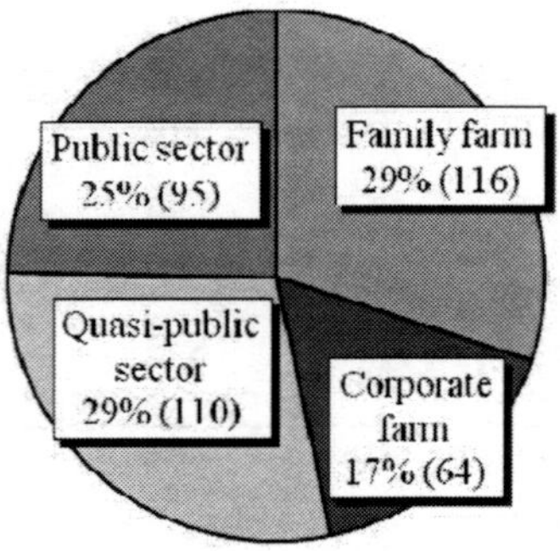

Note: sample size in parentheses

Figure 1: Proportion of providers.

Table 5: Types of providers and experience service.

No. of offered services

Items	Sample size	Provided services: Crop-wise Yes	No	test	Operation-wise Yes	No	test	Food culture Yes	No	test	Rural culture Yes	No	test
Family operation	116	2.1	2.8	***N	3.1	3.1	n.s.	1.8	1.7	n.s.	1.2	1.3	n.s.
Corporate operation	64	2.3	2.1	n.s.	3.1	3.1	n.s.	1.7	1.7	n.s.	1.4	1.2	n.s.
Quasi-public sector*	110	2.7	2.5	n.s.	3.2	3.1	n.s.	2.4	1.5	***N	1.8	1.1	***E
Public sector**	95	3.2	2.4	***N	3.0	3.1	n.s.	1.0	2.0	***N	0.7	1.5	***N
Educational dairy farms	93	1.9	2.8	***N	3.0	3.1	n.s.	1.3	1.8	***N	0.7	1.5	***N
Green tourism farms	24	3.0	2.5	+E	3.3	3.1	n.s.	2.6	1.7	**E	2.1	1.2	***E

Source: Same as Table 2.
Note 1: Test method: t test, E= equal variance, N= not equal variance and ***,**, *, + indicate 1%, 5%, 10%, 20% (reference), respectively and n.s.= not significant.
Note 2:*quasi-public sector= NGO, non-profit organizations abd agricultural cooperatives, etc.
**Public sector=governments, reserch institutes, colleges.

Operation-wise, there were no statistically significant differences among types of providers. Educational dairy farms provided fewer numbers of every type of service except operation-wise services (1% significance). Dairy farming is constrained in that only a single product is produced. Green tourism, or Japanese rural tourism, farms offered more services among food and rural experiences than non-green tourism farms with statistical significance.

In short, the public sector has the advantage in crop-wise experiences, while such an advantage was shown for the quasi-public sector in food and rural experiences. Family farms and educational dairy farms offered fewer numbers of services due to a limited variety of products. In addition, it is considered that the differences in the scope of the providers' network in each sector accounted for some of these differences in types of services.

As to facilities, first, with respect to Internet accessibility, having a URL was associated with providing a greater number of services (Table 6). Likewise, e-mail availability made a significant difference in the number of services offered. Therefore, it is safe to say that Internet availability is a necessary condition for the provision of educational services.

The number of offered services was significantly higher for food and rural culture experiences among those providers that charged for services, that targeted visitors nationwide, and that had a nearby accommodation facility than those whose services were offered free of charge, were limited to local visitors, and did not have accommodations nearby. These results indicate the value of attracting visitors from afar rather than limiting provision of services to the local population. Also overnight experience services combined with nearby accommodation facilities would be effective in promoting both experience services and rural tourism. Charging for services is undertaken to recover material costs used for experience services and opportunity cost for providers. Further, charging is an important step toward enhancing the recognition and value of these newly emerging services.

Table 6: Characteristics of providers and experience services.

No. of offered services

Items	Sample size	Provided services											
		Crop-wise			Operation-wise			Food culture			Rural culture		
		Yes	No	test	Yes	No	test	Yes	No	test	Yes	No	test
E-mail availability	234	2.8	2.2	***N	3.2	2.9	*E	1.9	1.4	**N	1.4	1.2	n.s.
Own URL	210	2.8	2.2	***N	3.2	3.0	+N	1.9	1.4	***N	1.5	1.1	***N
Charging for services	222	2.6	2.5	n.s.	3.1	3.1	n.s.	2.3	1.0	***N	1.7	0.7	***N
Partially charging	62	2.3	2.6	+E	3.3	3.1	n.s.	1.5	1.8	n.s.	1.2	1.3	n.s.
Targeted area: whole country	281	2.5	2.8	n.s.	3.2	2.9	+E	2.1	0.7	***N	1.6	0.5	***N
Availability of nearby accommodations	49	2.8	2.5	n.s.	3.6	3.0	**E	2.8	1.6	***N	2.2	1.2	***E
Utilization of renovated closed school	4	5.8	2.5	***E	4.8	3.1	**E	5.3	1.7	***E	3.5	1.3	***E
Availability of instructors	140	2.9	2.4	***E	3.3	3.0	*E	2.5	1.3	***N	1.8	1.0	***N
Availability of indoor	236	2.7	2.3	**E	3.3	2.9	**N	1.9	1.4	**N	1.4	1.1	**N
Availability of bathroom for the disabled	64	3.0	2.5	**E	3.2	3.1	n.s.	2.5	1.6	***E	2.1	1.1	***N

Source: Same as Table 2.

Note: Test method: t test, E= equal variance, N= not equal variance and ***,**, *, + indicate 1%, 5%, 10%, 20% (reference), respectively and n.s.= not significant.

Looking at other aspects, utilization of renovated once-closed schools, providing instructors for guiding visitors in experiencing aspects of farm life, having indoor programs, and installation of a bathroom for the disabled all work positively on an increase in the number of services.

In summary, all of the conditions described above work positively to increase the number of provided experience services related to both technical and institutional complementarity. We cannot ignore the role of the public sector in experience services related to agricultural production with tighter technical complementarity because this sector plays a role in networking of providers. On the other hand, the quasi-public sector, such as NGO, plays an active role in culture experiences with institutional complementarity by charging for these services and accommodating visitors from a wide area. Thus, these results indicate that educational services go beyond the scope of technical and institutional complementarity. Since the quasi-public sector has an affinity for food and rural culture experience services, we should especially recognize and utilize this sector in promotion of culture-related educational services.

5 Conclusions

This paper focused on educational services related to agriculture as joint products of agriculture and investigated the characteristics of providers of these services in Japan by incorporating technical and institutional complementarity. The main findings are as follows.

(1) Experience services are classified from the viewpoint of complementarity in that crop-wise and operation-wise experience services are defined by technical

complementarity with tight seasonality while food and rural culture experience services are defined by institutional complementarity with less seasonality.

(2) NGOs and non-profit corporations play a positive role in provision of food and rural culture experiences, which indicates that institutional complementarity of rural life aspects has an affinity with the quasi-public sector. This fact shows that the quasi-public sector will be able to promote the educational function in agriculture throughout a wider area because of the range of their networking capability than could be achieved within the range of original institutional complementarity, which was limited to farm households and rural community.

(3) Common issues that need improvement are Internet availability, charging for services and training programs aimed to help providers offer better service quality. Partnership with NGO and non-profit corporations and the public sector will be effective in easing the constraints of seasonality and increase services that can be accompanied by an overnight stay.

(4) Consequently, it is important to recognize the significance of institutional complementarity because it will widen our perspective for farm diversification and the promotion of the educational function. Further study is necessary on the relationship between cultural aspects and farm diversification.

References

[1] Hagedorn, K., Rethinking the Theory of Agricultural Change in an Institution of Sustainability Perspective, G. Van Huylenbroeck, W. Verbeke, L. Lauwers, I. Vanslembrouck, and M. D'Haese (eds.) *Importance of Policies and Institutions for Agriculture*, Academia Press: Gent, pp.33-56, 2003.

[2] OECD, *Multifunctionality: Towards an Analytical Framework*, OECD: Paris, 2001.

[3] OECD, *Multifunctionality: The Policy Implications*, OECD: Paris, 2003.

[4] Ohe, Y., On the Relationship between Multifunctionality and Hamlet Activities as a Rural Institution, *Jpn. J. Rural Econ.* Vol.8, pp.41-57, 2006.

[5] Ohe, Y., Evaluating Jointness of Multifunctional Agriculture: the Educational Function of Dairy Farming in Japan, Aravossis, K., C. A. Brebbia, E. Kakaras, and A. G. Kungolos (eds.) *Environmental Economics and Investment Assessment*, WIT Press: Southampton, pp.337-346, 2006.

[6] Japan Science Council, Evaluating Multifunctionality in Agriculture and Forestry, Connecting Global Environment and Human Life (Report), p.16, 2001.

www.witpress.com, ISSN 1743-3541 (on-line)

Green economies and green jobs: implications for South Africa

G. Nhamo
Centre for Corporate Citizenship, University of South Africa, South Africa

Abstract

The term 'green' is often (mis)used in a number of ways. For example, we talk and hear of green buildings, green financing, green economies, green wash, green accounting, green schools, green agriculture, green tourism, green procurement, green ITC, the list goes on. In this paper, we address twin, but closely linked, green terms: 'green economy' and 'green jobs' and their implications for an emerging economy like South Africa. The paper conceptualises the transition to green economies and how this impacts on the jobs market in South Africa.

Keywords: green economy, green jobs, South Africa.

1 Introduction

A landmark and comprehensive publication in the area of green economies and green jobs is a 2008 joint document titled *Green Jobs: Towards Decent Work in a Sustainable, Low-Carbon World* produced by the United Nations Environment Programme (UUNEP), International Labour Organisation (ILO), International Organization of Employers (IOE), International Trade Union Confederation (ITUC). The publishers claim that the publication is in large part as a result of climate change (CC) and the need to meet greenhouse gas (GHG) emission reduction targets under the UN Framework Convention on Climate Change (UNFCCC) and the Kyoto Protocol market based mechanism (UNEP *et al.* [1]). The need to urgently address CC has led to changing patterns of investment flows leading to investment into areas from renewable energy generation up to energy efficiency projects at the household and industrial level. Many green jobs have been created during the transition to green economies. For example, a

WIT Transactions on Ecology and the Environment, Vol 131, © 2010 WIT Press
www.witpress.com, ISSN 1743-3541 (on-line)
doi:10.2495/EEIA100221

project in Bangladesh, training local youths and women as certified solar technicians and as repair and maintenance specialists, aims to create some 100,000 jobs. In India, an initiative to replace inefficient biomass cooking stoves in nine million households with more advanced ones could create 150,000 jobs (UNEP *et al.* [1 p. vii]). Many more green jobs are projected into the future of which the South African economy must be party to.

2 Green economies

Green economies are so-called because they exhibit high levels of jobs created to address a myriad of environmental concerns. In an ideal state of affairs, a green economy is one that does not generate pollution or waste and is hyper-efficient in its use of energy, water, and materials. Green economies are also exemplarily in terms of the manner in which a deliberate and calculated move towards low carbon emissions is made, thereby creating low-carbon economies and those that we can consider to be climate change compliant. To achieve the low-carbon economy status, innovative ways of thinking are getting embedded in Low Carbon Growth Plans (LCGP), like South Africa's DEAT [2], through the Long Term Mitigation Strategy (LTMS) that needs to be complemented by a similar initiative possibly for a Long Term Adaptation Strategy and the National Climate change Response Strategy.

A transition to green economies implies four major scenarios in terms of jobs UNEP *et al.* [1 p. 3]:

1. In some cases, additional jobs will be created – as in the manufacturing of pollution-control devices added to existing production equipment.
2. Some employment will be substituted – as in shifting from fossil fuels to renewables, or from truck manufacturing to rail car manufacturing, or from landfilling and waste incineration to recycling.
3. Certain jobs may be eliminated without direct replacement – as when packaging materials are discouraged or banned and their production is discontinued.
4. It would appear that many existing jobs (especially such as plumbers, electricians, metal workers, and construction workers) will simply be transformed and redefined as day-to-day skill sets, work methods, and profiles are greened.

The four bullets highlighted above are special for South Africa. When the cost-benefit analysis of the green jobs is done for the country these facets need to be taken into consideration. It terms of skills development, the workers highlighted under bullet 4 will require 'greening' (Figure 1).

Chapple [3 p. 1] maintains that at basic level, a green economy is a clean energy economy comprising primarily four sectors: renewable energy (e.g., solar, wind, geothermal); green building and energy efficiency technology; energy-efficient infrastructure and transportation; and recycling and waste-to-energy. In terms of policy and legal frameworks, national governments must ensure that these result in the provision of funding of green projects, including infrastructure development where the private sectors lack UNEP *et al.* [1].

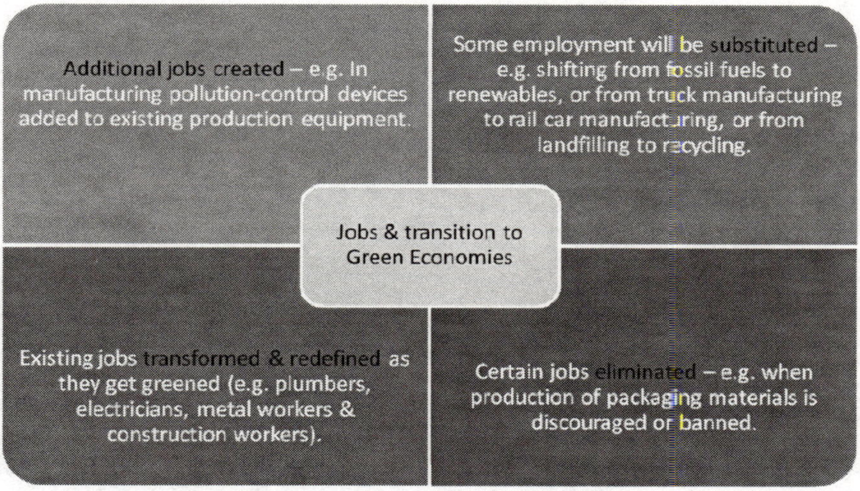

Figure 1: Green jobs and transition to green economies. Source: author.

Governments must create and maintain a level playing field for all actors and among major policies can be listed UNEP *et al.* [1 p. 5]:

- *Subsidies*: Phase out subsidies for environmentally harmful industries, and shift a portion or all of those funds to renewable energy, efficiency technologies, clean production methods, and public transit.
- *Carbon Markets*: Fix the current shortcomings inherent in carbon trading and Kyoto Protocol related innovations like the Clean Development Mechanism so that they can become reliable and adequate funding sources for green projects and employment.
- *Tax Reform*: Scale up eco-taxes, such as those adopted by a number of European countries, and replicate them as widely as possible. Eco-tax revenues can be used to lighten the tax burden falling on labour while discouraging polluting and carbon-intensive economic activities.
- *Targets and Mandates*: Ensure that regulatory tools are used to the fullest extent in the drive to develop greener technologies, products, and services – and thus green employment. This includes land-use policies, building codes, energy-efficiency standards (for appliances, vehicles, etc.), and targets for renewable energy production.
- *Energy Alternatives*: Adopt innovative policies to overcome barriers to renewable energy development, including feed-in laws that secure access to the electrical grid at guaranteed prices.
- *Product Take back*: Adopt "extended producer responsibility" laws (requiring companies to take back products at the end of their useful life) for all types of products.
- *Eco-Labelling*: Adopt eco-labels for all consumer products to ensure that consumers have access to information needed for responsible

purchasing decisions (and hence encouraging manufacturers to design and market more eco-friendly products).

- *R&D Budgets*: Reduce support for nuclear power and fossil fuels and provide greater funding for renewable energy and efficiency technologies.
- *International Aid*: Reorient the priorities of national and multilateral development assistance agencies as well as export credit agencies away from fossil fuels and large-scale hydropower projects toward greener alternatives.

HSBC Global Research cited in WWF [4 p. 6] notes that a €2.2 trillion worth of global stimulus packages worldwide has been instituted since 2008. Of this, close to €300 billion addresses climate policy objectives. Europe accounts for €42 billion of this, or about 13% of the global total. European stimulus packages provide a boost to Carbon Capture and Storage (CCS) research and low-carbon vehicles and, to some extent, energy efficiency of buildings. The European Commission's Economic Recovery Plan finalised in June 2009 includes, the green cars (€5 billion) initiative, energy-efficient buildings, €1 billion for research and development, and "factories of the future" (€1.2 billion for R&D). Close to 60% of the funds were green.

3 Towards a green jobs definition

Defining complex issues is always complex. However, in order to provide a solid operational foundation in as far as green jobs perspectives are concerned in South Africa, an attempt will be made to present a number of definitions for such. Green Jobs are defined by the UNEP *et al.* [1 p. 3] as:

> Work in agricultural, manufacturing, research and development (R&D), administrative, and service activities that contribute substantially to preserving or restoring environmental quality. Specifically, but not exclusively, this includes jobs that help to protect ecosystems and biodiversity; reduce energy, materials, and water consumption through high efficiency strategies; de-carbonize the economy; and minimize or altogether avoid generation of all forms of waste and pollution.

What is of interest here is the fact that the authors identify six key sectors that have a bearing on environmental preservation and restoration. Effectively, these key sectors play a critical role in terms of generation Green Jobs opportunities. Green jobs must be decent work, offering adequate wages, safe working conditions, job security, reasonable career prospects, and worker rights. The movement toward Green Economies is viewed as having more benefits in terms of job opportunities than the disadvantages UNEP *et al.* [1]. Coincidentally the above definition of green jobs has been adopted by the South African government as it moves towards the development of a national green jobs strategy planned to be finalised in 2010.

A study in the USA by Pollin *et al.* [5] focusing on six infrastructure investment strategies looking at retrofitting buildings to improve energy efficiency, expanding mass transit and freight rail, constructing smart electrical grid transmission systems, expanding production of wind power, expanding production of solar power and next-generation biofuels development reveals that a short term $100 billion green economic recovery could create two million jobs opportunities across board. No definition of green jobs was presented.

Japan started communication around green jobs as early as 1994. Making reference to the green jobs potential in Japan, McClenahen [6] maintains that Japan had the potential to create 18,000 jobs in the constructions sector and 17,000 in environmental research alone in the period 1994–2000. The predictions were based on information from Japan's Environmental Protection Agency that assumed a constant annual growth rate of 8% and an annual investment of $959 million in twin sectors identified.

Global Insight [7] was sponsored by the United States Conference of Mayors Climate Protection Centre to consider the USA Metro Economies focusing on Current and Potential Green Jobs in the USA Economy. The time frame looked at 30 years covering 2008 to 2038. The industry scope focused on renewable power generation, residential and commercial retrofitting as well as renewable transportation fuels. In the study, an attempt was made to define green jobs a definition of which is provided here. Green bobs were defined as:

> Any activity that generates electricity using renewables or nuclear fuels, agriculture jobs supplying corn or soy for transportation fuel, manufacturing jobs producing goods used in renewable power generation, equipment dealers and wholesalers specializing in renewable energy or energy-efficiency products, construction and installation of energy and pollution management systems, government administration of environmental programs, and supporting jobs in the engineering, legal, research and consulting fields (Global Insight [7 p. 5]).

In terms of green job opportunities, Global Insight noted a figure of 751,000 as of 2006. From Global Insight's perspective, this figure amounted to less than 0.5% of total current jobs. Of the 751,000 green jobs, an estimated 85% was generated in metropolitan areas. For the 30 year horizon under consideration, an estimated 4.2 million new green jobs potential existed. A further breakdown on the spread of the green jobs potential was made as follows:

- Renewable Power Generation: 1.2 million green jobs;
- Residential and Commercial Retrofitting: 81,000 jobs – 36,000 in the residential sector and 45,000 in commercial; and
- Alternative Transportation Fuels: 1.5 million.

Ashkin [8 p. 19] establishes that one of the best definitions presented concerning green jobs is one by Carl Pope, an executive director of the Sierra Club who said: "A Green job has to do something useful for people, and it has to be helpful to, or at least not damaging to, the environment'. The author sees green jobs as helping to create a powerful retaliation against climate change, with social and economic benefits that are multifaceted. The green jobs are local, and cannot be outsourced overseas. Aspects like installing solar panels need on-site labour whist retrofitting an office building is not something that can be done by remotely. The green jobs have an additional advantage in that they promise a middleclass wage, many do not require a college education, and are going to be in high demand in urban, as well as rural communities, an aspect that sits well with South Africa's poverty alleviation and jobs creation agenda.

A study by Bezdek [9] commissioned by American Solar Energy Society to look into 'Green Collar Jobs in the U.S. and Colorado: Economic Drivers for the 21st Century' is another piece of work. The time line covered 2006 to 2030 with a focus on assessing direct job creation in renewable energy and energy efficiency. The study also looked at indirect green jobs potential all other sectors of the USA economy. Another definition on green jobs was conceptualised as direct job creation estimates that include employees working in one of the major renewable energy technologies like wind and solar power as well as employees working in industries that are entirely in the energy efficiency sector as well as a fraction of employees working in industries partially involved in energy efficiency.

To this end, green jobs opportunities in the renewable energy industry would include workers in the wind, solar, hydroelectric, geothermal, biomass, fuel cell, and hydrogen technology industries. Green jobs opportunities in the energy efficiency sector would include workers in areas like recycling, reuse or remanufacturing and a share of employees in industries that are partially in the energy efficiency sector like household appliances, vehicle production, and construction. Indirect green jobs potential also exists in sectors like restaurant staff or dry cleaners employed due to demand from directly created green jobs Bezdek [9].

The greatest percentage of jobs created in renewable energy and energy efficiency are, however, conventional jobs for accountants, engineers, computer analysts, clerks, factory workers, truck drivers and mechanics of which many employees in this group might not even realize that they owe their livelihood to renewable energy and energy efficiency. The green jobs potential in the wind sector can best be illustrated by considering cases from Denmark, Germany and Spain Boettcher *et al.* cited in (Kammen and Engel [10 p. 8]).

Bezdek [9] notes that in 2007, renewable energy was projected to be responsible for roughly 500,000 jobs, of which about half were indirect. More than 70% of the half a million jobs were in the biomass sector, with the largest share contributed by the ethanol and biomass power sub-sector. In terms of growth, the renewable energy sector was estimated to have grown 15% in two years (2006–2007), more than seven times as fast as the total U.S. economy growth. The energy efficiency sub-sector contributed about 8.5 million jobs with about 3.7 million of those being direct jobs in 2007. More than 36% of the jobs were in the recycling, reuse and remanufacturing sector. Overall, Bezdek [9] projected total jobs related to renewable energy and energy efficiency by 2030 under three scenarios.

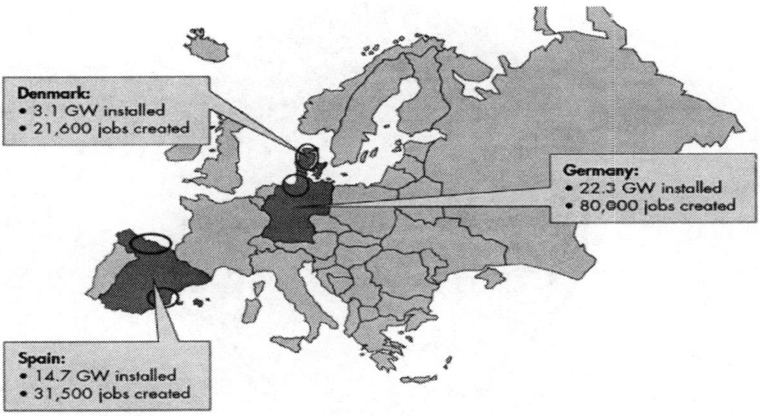

Figure 2: Installed capacity of wind energy and jobs created. Source: Boettcher *et al.* (cited in Kammen and Engel [10 p. 8]).

The work also proposes that the amount of jobs could go up depending on the aggressiveness of the key USA actors in promoting renewable energy and energy efficiency. Such a stance would lead to nearly six times as many jobs in 2030. This might not be a difficult task to achieve given that President Obama has firmly placed renewable energy and energy efficiency initiative on top of the USA government agenda. Another earlier study to project green jobs from the clean energy industry Kammen *et al.* [11] is available for further reference.

South Africa talks of the concept environmental jobs McCutcheon [12], DoPW [13]. Environmental jobs must be seen as a component of Green Jobs. In this regard, the South African EPWPII identified environmental and other jobs from other sectors that include infrastructure, social and non-state. Forstater [14] sees a supportive role played by public sector employment programmes in creating green jobs and sustaining the environment. Hence a diagrammatic representation to show and differentiate between jobs is presented in Figure 3.

It is important to note that green jobs are an integral component of the greater jobs market. It is also critical to note that from Figure 3 green jobs are environmental jobs and not vice versa. In addition, at the core of green jobs are jobs oriented towards addressing CC, chief among them, those from energy efficiency initiatives, alternative energy and any other addressing the need to move economies towards low carbon emission.

4 Green jobs on the South African agenda

Up until recently, there has not been a clear-cut national policy that addresses the concept of green jobs in South Africa. A number of national documents and high level politicians and business leaders have made *ad hoc* reference to the concept. Early work towards the conceptualisation of green jobs was done by the Department of Trade and Industry (DTI) in 2006. In a document entitled

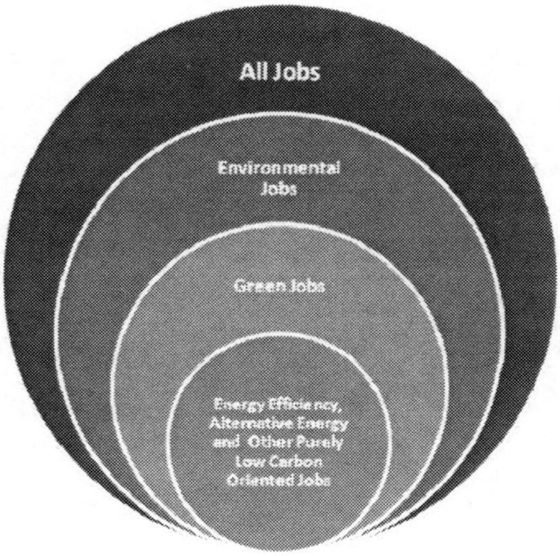

Figure 3: Conceptualisation of the jobs market. Source: author.

'Programme to Stimulate Growth in the South African Environmental Goods and Services Industry: The Green Jobs Concept', the DTI made reference to green jobs although no definition was provided. The DTI in collaboration with industry's intention was to evaluate the strengths and weaknesses of the Environmental Goods and Services (EGS) industry with a view to develop such a programme for South Africa (Forstater [14], DST [15]). The EGS industry was realised to be among the fastest growing industries globally, especially in line with environmental and climate change challenges. The DTI also noted that unlike other industrial sectors, the growth of the EGS industry is largely driven by the activities and policies of governments (DST [15]). Drawing from the UK, OECD and other global examples and classifications, the DTI list six sub-sectors as having potential for export growth and investment leading to the creation of green jobs in the country (DST [15 p. 3-5]) namely:

1. Environmental Services and Consultancy;
2. Water and wastewater treatment (to be prioritised);
3. Waste Management, Recovery and Recycling (to be prioritised);
4. Air Pollution Control;
5. Emergency response/clean-up; and
6. Clean Development Mechanism (CDM) and ?Energy Efficiency

Given the level of understanding pertaining to EGS in 2006, one cannot therefore blame the DTI for the strategic priorities made then. In fact, the DTI went further and prepared a road map for the recycling industry in South Africa (DTI [17]) that was published in January 2009. However, with the current government thinking in terms of the urgent need to address climate change, DTI and other government departments' priorities are likely to change and focus more on energy to include energy mix and energy efficiency. From the recycling industry road map, DTI indicated that its preferred approach to growing the EGS industry sector was to focus on and promote sub-sectors like cleaner production, energy co-generation and recycling. A full operating capacity from the recycling industry was estimated to "increase the number of direct employment opportunities by between 18,000 and 100,000 to offer 149,459 direct employment opportunities" (DTI [17 p. 1]) and up to 190,000 jobs by 2015.

A draft discussion document towards 'The National Climate Change Response Policy' for South Africa presented during the National Climate Change Summit mentions the concept of green jobs twice. It has this to note concerning green jobs:

> The In the USA and the UK and on the agenda of the G20 is the concept of a "green" or 'low carbon' path to economic recovery. Indeed, our President in the State of the Nations Address at the opening of Parliament in February, referred to the creation of '*green jobs*' (emphasis is the author's) (DEAT [18 p. 12]). … South Africa's current energy supply crisis has created an opportunity for a thorough going transformation in the way that our economy and society is powered. Additionally, '*green jobs*' and economic initiatives are being proposed worldwide as a key element of an economic recovery strategy (DEAT [18 p. 18]).

The 'Framework for South Africa's Response to the International Economic Crisis of February 2009 mentions 'green jobs' once. The concept is mentioned within the context of climate change, technology and industrial development. Item 3.6 in the Framework records that:

> The parties recognise the opportunities in industries that combat the negative effects of climate change and believe that South Africa should develop strong capacity in these green technologies and industries. Accordingly it is agreed to develop incentives for investment in a programme to create large numbers of *'green jobs'*, namely employment in industries and facilities that are designed to mitigate the effects of climate change The Presidency [19 p.11].

The Framework for South Africa's Response to the International Economic Crisis set aside 787 billion Rand, a figure similar to President Obama's US$787 billion. President Obama's $787 billion Stimulus of 2008 was projected to create about 1.4 million green jobs in addition to 5 million jobs projected from another $150 billion set aside for 2009–2018/19 for the clean energy sector Winkler [20]. Other things being equal, a clever guess could be that the 787 billion Rand stimulus for South Africa could generate a similar number green jobs (Creamer [21]).

In a budget vote speech made by the Minister for Agriculture, Forestry and Fisheries to Parliament in June 2009, the concepts of green jobs was alluded to. Minister Joemat-Petersson indicated that the government needs "to develop appropriate responses and invest in cleaner production methodologies, create *green job* opportunities, promote innovation and apply scientific technology in the production processes" (Joemat-Petersson [22 p. 5]).

The concept of environmental jobs has been formally recorded in the Expanded Public Works Programme (EPWP) (DoPW [13]). The EPWP is one of the key poverty and hunger alleviation programmes that the government has been driving vigorously. Significant progress has been made towards achieving the EPWP target of creating more than a million net new work opportunities over the five-year period of the programme. The EPWP recorded up to 281,626 jobs (38% of the targeted 750,000 work opportunities) as having been created through the Infrastructure Sector Programme (ISP) (The Presidency [23 p. 9]).

The EPWP is run under four distinctive sectors namely: environment and cultural, social, economic and infrastructure sectors. These sectors comprise a number of government departments with one department nominated to lead each sector (DoPW [13]). The former Department of Environmental Affairs and Tourism (DEAT) was nominated to lead the Environmental and Cultural Sector consisting of three other departments then: Water Affairs and Forestry, Arts and Culture as well as Agriculture (DoPW [13]).

The job opportunities were highlighted to be created in five key programme areas including Sustainable Land Based Livelihoods, Working for the Coast, People and Parks, Working for Tourism and Working on Waste. The sustainable land based livelihoods sub-sector addressed job opportunities under programmes that included Working for the Land, Working for Water, Working for Wetlands and Working on Fire (DoPW [13]).

The President of the Republic of South Africa, Jacob Zuma announced an estimated 500,000 jobs had to be created by December 2009. This was quiet a bold move given that the world over every economy was battling with the financial crisis and job losses. The road map informing the proclamation by the President is enshrined in the Expanded Public Works Programme II (EPWPII) whose job projections and details discussed in much depth in the next sections. The total full time job equivalent (including those from the environmental sector) is presented in Figure 4.

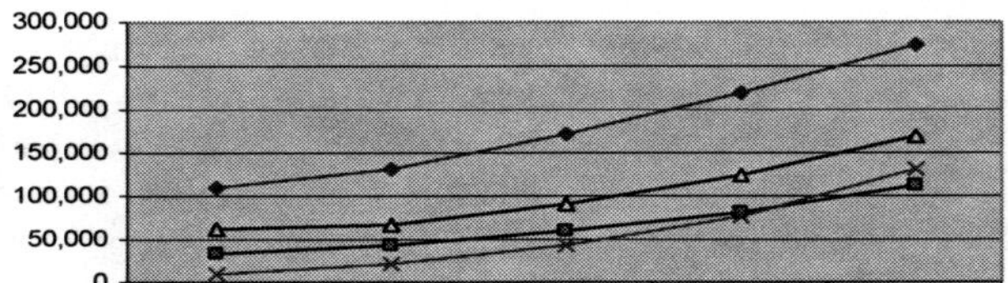

	2009-2010	2010-2011	2011-2012	2012-2013	2013-2014
Infrastructure	108,696	130,435	170,435	220,000	273,913
Environmental	32,609	41,739	59,130	81,304	110,870
Social	60,870	67,826	90,435	124,348	169,565
Non-State	8,696	20,870	41,739	76,522	130,435

Figure 4: Total full time equivalent (job) by sector. Source: author.

5 Future of green jobs in South Africa

The success of the green jobs initiative in South Africa is dependent upon a wide range of factors and actors. Organised government, as well as the private sector must play a pivotal role supported by the NGO and donor community. Changes in the decisions, practices and behaviours of managers, workers and consumers will also add value to this transition process.

A green South African economy will create demand for workers, many of them in skilled trades or professions, and filling these positions will require adequate training programs. At the cutting edge of technology development for wind turbine or solar PV design, for instance, specialization has progressed to the point where universities need to consider offering entirely new study fields and majors. Several countries have reported that a "skills gap" already exists between available workers and the needs of green industries (UNEP *et al.* [1 p. 25]):

- A 2007 survey of Germany's renewables industry concludes that companies in this field are suffering from a shortage of qualified employees, and especially those needed in knowledge-intensive positions.
- The Confederation of British Industry has expressed concern that sectors going green are struggling to find technical specialists, designers, engineers, and electricians.
- In the United States, the National Renewable Energy Laboratory has identified a shortage of skills and training as a leading barrier to renewable energy and energy-efficiency growth.

The challenges highlighted above are also present in South Africa. As a high GHG emitter, the country must move fast to engage with mechanisms leading to transition to a green and low carbon economy. The upcoming national green jobs strategy and other planned policy and legal frameworks must be finalised without delay. The greatest challenge will still remain the skills gap.

References

[1] UNEP/ILO/IOE/ITUC. Green Jobs: Towards Decent Work in a Sustainable, Low-Carbon World. 2008. Nairobi, UNEP/ILO/IOE/ITUC.

[2] DEAT. Long Term Mitigation Scenarios Strategic Options for South Africa. 2007. Pretoria, Government Printer.

[3] Chapple, K. Defining the green economy: A primer on green economic development. 2008. Berkeley, Centre for Community Innovation.

[4] WWF. Low Carbon Jobs for Europe: Current opportunities and future prospects. 2009. Brussels, WWF European Policy Office.

[5] Pollin, R.; Garrett-Peltier, H.; Heintz, J.; Scharber, H. Green recovery: A program to create good jobs and start building a low-carbon economy. 2008. Massachusetts Amherst, Center for American Progress.

[6] McClenahen, J. Green jobs in Japan. Industry Week, 89-90. 1994.

[7] Global Insight. U.S. Metro Economics: Current and potential green jobs in the U.S. economy. 2008. Lexington, Global Insight.
[8] Ashkin, S. The green revolution: Not your great-grandfather's revolution. 2007. New York, The Green Scene.
[9] Bezdek, R. Renewable energy and energy efficiency: Economic drivers for the 21st Century. 2007. New York, American Solar Energy Society.
[10] Kammen, D. M.; Engel, D. Green jobs and the clean energy economy. 2009. Berkeley, Copenhagen Climate Council.
[11] Kammen, D. M.; Kapadia, K.; Fripp, M. Putting renewables to work: How many jobs can the clean energy industry generate? 2006. Berkeley, University of California.
[12] McCutcheon, R. T. Employment Generation in Public Works: Recent South African Experience. *Construction Management and Economics* **2001**, 19, 275-284.
[13] DoPW. Environment and Culture Sector Plan. 2004. Pretoria, Department of Public Works.
[14] Forstater, M. Green Jobs: Public Service Employment and Environmental Sustainability. *Challenge* **2006**, 49, 58-72.
[15] DST. Green jobs and industries for South Africa (Green-JISA). 2009. Pretoria, Department of Science and Technology.
[16] DTI. Programme to stimulate growth in the South African environmental goods and services industry: The green jobs concept. 2009. Pretoria, Department of Trade and Industry.
[17] DTI. Proposed road map for the recycling industry: A recycling industry body (RIB) with relevant working groups. 2009. Pretoria, Department of Trade and Industry.
[18] DEAT. The National Climate Change Response Policy: Discussion Document for the 2009 National Climate Change Response Policy development Summit, Gallagher Convention Centre, Midrand, 3-6 March 2009. 2009. Pretoria, Department of Environmental Affairs and Tourism.
[19] The Presidency. Framework for South Africa's Response to the International Economic Crisis - 19 February 2009. 2009. Pretoria, The Presidency.
[20] Winkler, H. Working for Energy. http://www.engineeringnews.co.za/article/working-for-energy-2009-02-20 (accessed 13 September 2009).
[21] Creamer, T. 'Green jobs' plan in SA stimulus framework. http://www.engineeringnews.co.za/article/green-jobs-plan-included-in-sa-stimulus-framework-2009-02-23 (accessed 30 August 2009).
[22] Joemat-Petersson. SA: Joemat-Petersson: Budget vote speech by the Minister for Agriculture, Forestry and Fisheries to Parliament (17/06/2009). 2009. Pretoria, Ministry of Agriculture, Forestry and Fisheries.
[23] The Presidency. ASGISA Annual Report 2007. 2007. Johannesburg, The Presidency.

Author Index

Brownfields IV

Prevention, Assessment, Rehabilitation and Development of Brownfield Sites

Edited by: ***E. BERIATOS**, University of Thessaly, Greece and **C.A. BREBBIA**, Wessex Institute of Technology, UK*

Containing papers presented at the Fourth International Conference on Prevention, Assessment, Rehabilitation and Development of Brownfield Sites, this book discusses the problems facing public and private sectors, and the engineering and scientific communities, in terms of the land available for development purposes. The Conference looked at long-term plans for the productive re-use of properties that have been abandoned or lie idle, in order to satisfy current needs without compromising the ability of future generations to meet their own requirements.

Brownfield redevelopment is not solely an environmental issue, as it requires the involvement of financial, regulatory and community interests; which makes the whole process complicated. This is the reason why lending institutions, investors and real estate developers are cautious when dealing with brownfield redevelopment. However, these entities have become more tolerant of potential exposure as they become more knowledgeable of the risk involved. Given the economic and social benefits of brownfield redevelopment, there is a need for guidance on a process of ensuring the acceptability and therefore viability of such redevelopment. The preparation of the guidance requires further research as well as the sharing of information, lessons and experience among experts in this field.

This book will be of particular interest to practitioners and those businessmen in industry and commerce as well as those in research organisations interested in the problems facing the prevention, assessment, rehabilitation and development of brownfields. The papers include: Risk Assessment and Management; Case Studies; Development Issues; Community and Public Involvement; Rehabilitation of Brownfields; Remediation Studies and Technologies.

WIT Transactions on Ecology and the Environment, Vol 107
ISBN: 978-1-84564-105-4 eISBN: 978-1-84564-309-6
2008 288pp £95.00